# Prospecting With Old E-Logs

# Prospecting With Old E-Logs

Rollyn W. Frank

SCHLUMBERGER • EDUCATIONAL • SERVICES

Printed in the United States of America
Order No. SMP-7013

# Preface

The term "old logs" most often refers to electrical logs, MicroLogs, Laterologs, Microlaterologs, and neutron logs recorded before 1958.

These logs make up almost 45% of all existing logs, and skill in their use and an understanding of their limitations are essential if they are to be used successfully in prospecting. This book is dedicated to this difficult and often frustrating task.

Rollyn W. Frank

# Charts and Nomograms

Throughout this book, I refer to various charts or nomograms. Charts and nomograms for the first 10 chapters are in Appendix A: *Log Interpretation Charts for Old Electric Logs*. Appendix B contains the charts for the formation tester described in Chapter 11. These charts are reproduced directly from old interpretation documents.

R.W.F.

# Contents



*(continued on next page)*

# Contents

*(continued)*

# Contents

*(continued)*



*(continued on next page)*

# Contents

*(continued)*

# 1 Historical Background

Between 1912 and 1926 Conrad Schlumberger, a French physicist, and his brother Marcel, a mechanical engineer, invented a mineral-prospecting technique based on electrical measurements taken at the surface of the earth. Their successful endeavor led first to "electrical coring" and then, in 1927, to the first "electric log."

In France's Pechelbronn field on September 5, 1927, resistivities were measured in a borehole and recorded on a graph called a "log." Although this log had only one curve, it was the first time a geological event had been related to an electrical measurement. The first electric log in the United States was recorded just two short years later in California on August 27, 1929. That log is shown in Fig. 1.

Another far-reaching discovery was made in 1931 — a spontaneous potential (SP) that existed in the borehole and could be measured. The curve from this measurement reflected every permeable layer and was an invaluable complement to the resistivity curve.

The efficiency of electrical logging was substantially enhanced by this discovery because although oil- and gas-bearing strata generally produced peaks in the resistivity curve, compact low-porosity beds gave peaks, also. In most cases the SP curve resolved the ambiguity; a resistivity peak without an SP anomaly most likely meant a nonpermeable, dry horizon.

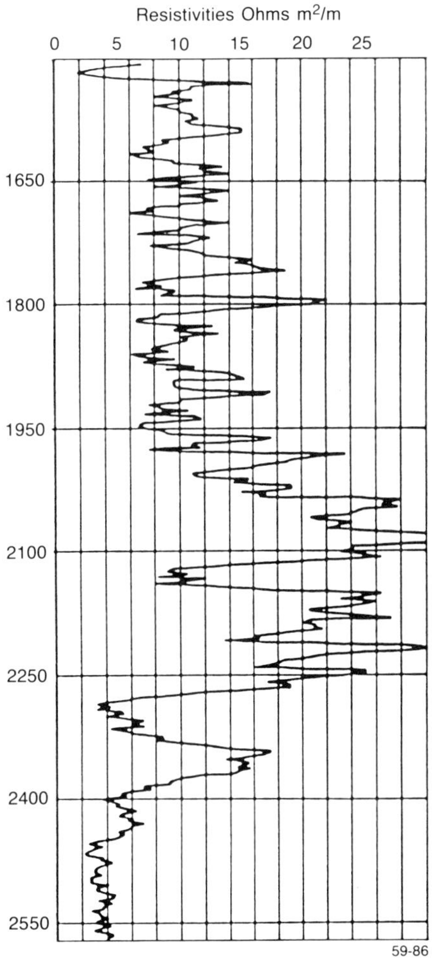


Fig. 1 First electrical coring diagram in the United States.

Fig. 2 Early logging truck.

The 1930's saw many changes in electrical logging. Experiments with various normal and lateral spacings were made to overcome the effects of mud invasion, which prevented an accurate measurement of true resistivity. A second normal curve was introduced in 1938 — another significant improvement. Thereafter, and for years to come, standard equipment included the short normal, the long normal, the lateral, and the SP. Fig. 2 shows an early logging truck, and Fig. 3 is a typical old electric log.

The following tables list other interesting information on the development of well logging and logging companies.

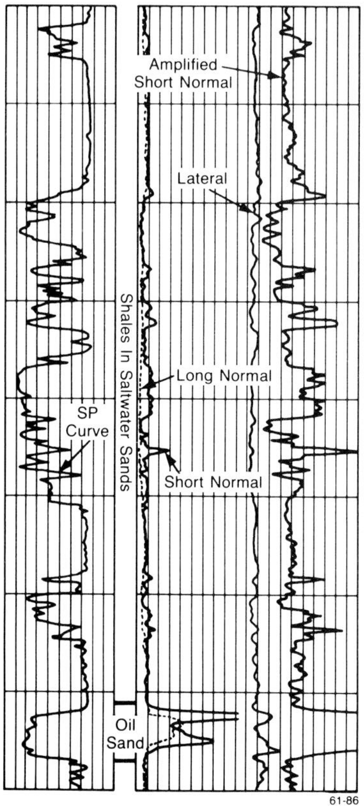


Fig. 3 Typical electrical log.

**Table 1**
Key Developments

| | |
|---|---|
| 1931 | Sidewall sampler |
| 1932 | Teleclinometer to measure borehole drift |
| 1933 | Temperature log |
| 1936 | SP dipmeter |
| 1942 | Archie's laws - first quantitative interpretation |
| 1945 | Neutron log |
| 1947 | Induction log |
| 1948 | MicroLog |
| 1948 | Determination of $R_w$ from the SP |
| 1951 | Laterolog and Microlaterolog |

**Table 2**
Early Logging Companies

| | |
|---|---|
| Schlumberger | 1927 |
| Halliburton | 1938 (Welex 1957) |
| Lane Wells | 1939 (Dresser-Atlas 1968) |
| Birdwell | 1948 |
| Elgen | 1953 (Lane Wells 1959) |
| PGAC | 1954 (Dresser-Atlas 1968) |

Halliburton became Welex in 1957. Dresser is an amalgamation of Elgen, Lane Wells, and Perforating Guns Atlas Corporation (PGAC).

Birdwell's early logs were mostly recordings of point electrode devices. They didn't offer the electric log until 1950.

During World War II, strategic raw materials such as nonferrous materials, tubular goods, and rubber were rationed by the War Production Board. The massive slowdown in drilling during this period forced Lane Wells to become inactive. Schlumberger and Halliburton continued business on a limited basis.

**Table 3**
Trade Names

| Company | Old E-Log | Focused Logs | Microdevices |
|---|---|---|---|
| Schlumberger | Electrical Survey | Laterolog<br>Microlaterolog | MicroLog |
| Halliburton | Electric Log | Guard Log<br>FoRxo Log | Contact Log |
| Lane Wells | Electrolog | Laterolog<br>Microlaterolog | Minilog |
| Elgen | Electronic Log | Laterolog<br>Microlaterolog | Permalog |
| PGAC | Electrolog | Laterolog<br>Microlaterolog | Micro Survey |

**Table 4**
Recorded Curves
Old E-Logs

| Company | Shallow | Medium | Deep |
|---|---|---|---|
| Schlumberger | 16-in. Nor.<br>10-in. Nor.<br>32-in. LS Lat.<br>37-in. LS Lat. | 64-in. Nor.<br>9-ft Lat.<br>9-ft Lat. | 18-ft 8-in. Lat.<br>15-ft Lat.<br>24-ft Lat. |
| Halliburton | 18-in. Nor. | 9-ft Lat. | 16-ft Lat. |
| Lane Wells | 16-in. Nor. | 40-in. Nor.<br>18-ft 8-in. Lat. | 16-ft Lat. |
| Elgen | 16-in. Nor. | 64-in. Nor. | 18-ft 8-in. Lat |
| PGAC | 16-in. Nor.<br>38-in. Nor. | 64-in. Nor. | 18-ft 8-in. Lat. |

LS = Limestone
Lat. = Lateral
Nor. = Normal

An understanding of the different normal and lateral devices recorded by each logging company is essential in the interpretation. Always check the log heading.

Presentations differed in various parts of the United States. For example, it was common to use the 10-in. normal and limestone lateral in west Texas carbonates, and California log presentations often varied considerably from those in other areas.

# 2 Fundamentals of Electrical Logging

Electrical logging consisted essentially of recording the resistivities of the subsurface formations and the spontaneous potentials generated in the boreholes. Presented to the petroleum industry more than 50 years ago, this important service was accepted as one of the most efficient tools in the search for, and the production of, oil and gas.

With electrical logging, parameters measured in situ by appropriate bottomhole instruments were continuously recorded at the surface. Measurements were performed only in the uncased portions of the boreholes.

It became a general practice, when a hole was drilled or at intervals during the drilling, to run an electrical survey for the purpose of quickly securing a complete record of the formations penetrated. This recording was of immediate value for the geological correlation of the strata and also for the detection and evaluation of possible productive horizons. The information obtained from the electrical logs was sometimes supplemented by sidewall samples of the formation or by other types of investigations performed in boreholes (deviation surveys, dipmeter surveys, temperature surveys, etc.).

Several different kinds of resistivity curves or logs, in addition to the SP curve, were recorded in the boreholes. These logs, obtained by using different resistivity measuring devices, are called conventional electrical logs (normal and lateral), Laterolog, MicroLog, etc. A typical electrical log recorded with the conventional devices is illustrated in Fig. 4. The left track of the log contains the SP curve, and the right tracks contain the resistivity curves. Each curve shows a sequence of peaks, deflections, and flat zones corresponding to the different formations traversed by the borehole. As will be seen in the following chapters, logs recorded with other types of resistivity measuring devices have a similar general appearance, although the corresponding devices differ basically in their principles and in their performances.

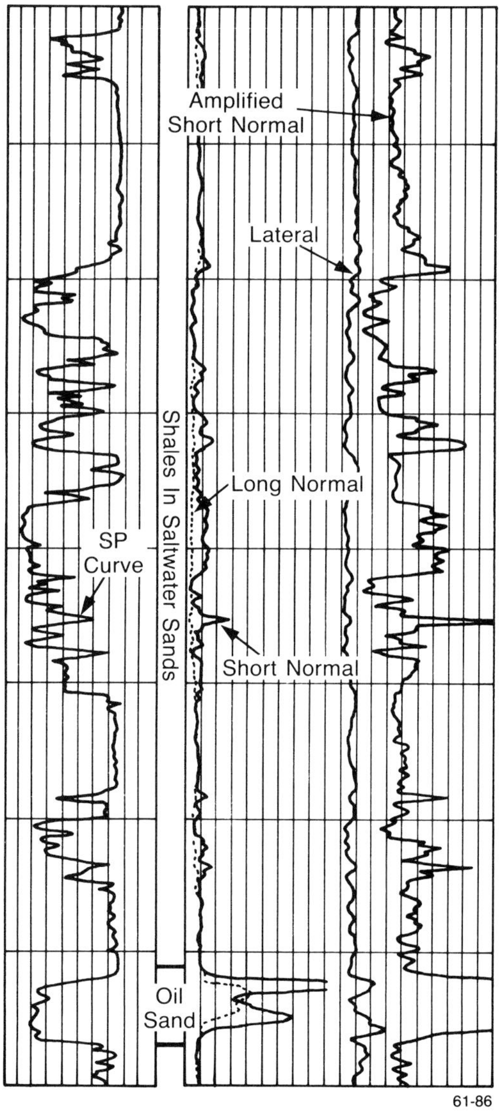

Fig. 4 Typical electrical log.

## RESISTIVITY METHODS

Resistivity methods are numerous, and it is convenient to classify them according to the corresponding types of subsurface measuring devices; i.e., macrodevices or microdevices.

### Macrodevices

The macrodevices investigate rather large volumes of material around the borehole, from about 10 to 100 cu ft and more. Their measurements are used for definition of formations, for correlations, and for qualitative and quantitative analyses of reservoirs in terms of fluid saturation and porosity. This group includes the conventional resistivity devices and the resistivity devices with focusing systems.

When using the conventional electrical log (normal and lateral), the path of the current used for the measurement has no restriction other than the location of the electrodes between which the current flows. Accordingly, the measurements are affected by the bed located at the level of the device, the mud column, and, if the bed is thin, the formations above and below it. The relative effect of the mud column and of the adjacent formations depends upon the physical configuration of the electrode system and the geometry and the respective resistivities of the media involved in the measurements. In formations of moderate resistivities, such as sands and shales, this effect is not too important; the conventional devices are generally suitable for a correct investigation except when the beds are very thin. The results are usually not so good in hard-rock countries where the resistivity contrasts between different beds and between the formations and the mud are high.

The devices that have focusing systems, such as the Laterolog, force the current to flow within a horizontal slice of formation of limited thickness. As a result, the effects of the mud column and of the formations above and below the limits of the horizontal slice are appreciably minimized and may be negligible in many cases. The focusing devices are better for investigating thin beds; the Laterolog is chiefly appropriate to hard formations, particularly in high-salinity muds.

### Microdevices

These pad-type devices investigate a very small volume of material (a few cubic inches) behind and close to the wall of the borehole, and the effect of the mud column on the measurement is practically eliminated. The delineation of the beds is, accordingly, much more accurate and more detailed than with the macrodevices, even those using focusing systems.

With the MicroLog system, the measurements are quite affected by the mudcakes, which makes it possible to detect permeable beds and define their boundaries exactly.

The Microlaterolog tool has a focusing system that minimizes the effect of the mudcake, or even eliminates it when the mudcakes are thin enough. Under these conditions the measurements give values close to the resistivity of the formations flushed by mud filtrate behind the wall of the hole. This provides an approach to determine the formation factor and, therefore, porosity.

When the porosity of the formation is sufficiently high, the formation factor and the porosity can also be determined by using MicroLog curves and appropriate correction charts.

## FORMATION RESISTIVITIES

The resistivities of the formations are important clues to their probable lithology and fluid content. Earth formations conduct electric current only by means of the mineralized water they contain. The minerals that constitute the solid parts of the strata are insulators when absolutely dry. In a similar manner, any pure oil or gas in the formations is electrically nonconductive. The few exceptions to this are metallic sulfides, graphite, etc., which conduct electricity like metals. Formations of lowest resistivity have increases in both the amount of water and its mineralization. Other important factors in formation resistivity are the shape and the interconnection of the pore spaces occupied by the water. These depend primarily on the lithology of the formation.

### Unit of Resistivity

Laterologs and electrical logs record the formation resistivity, and the unit is the ohm-meter (ohm-m). When a uniform electrical current is sent through a 1-m cube in a direction parallel to any edge, the resistance in ohms is equal to the resistivity of the substance. The same unit is also commonly designated ohm-meter squared per meter (ohm-$m^2$/m), which gives a good practical range of numerical values for the usual resistivities encountered in the earth, from 1 ohm-m or less to several thousand.

### Relation of Water Resistivity to Salinity and Temperature

The resistivity of a formation depends upon the resistivity of the water contained in its pores, which, in turn, is a function of the concentration of salts in that water.

In electrical log interpretation the resistivity of a formation water quite often has to be deduced from a knowledge of its salt content. Also, most interpretations involve comparison of different resistivity measurements, such as the resistivities of the drilling mud and of the formation water. In making such determinations, the following laws of electrolytic conductance should be kept in mind.

1. The conductivity of an electrolyte increases in proportion to the amount of chemicals in solution; the resistivity decreases with increasing salinity. Formation waters and drilling muds usually contain several chemicals in solution that differ in their ability to conduct electrical current. Since the predominant chemical is sodium

chloride (NaCl), the resistivity graph for NaCl solutions (Fig. 5) can be used to convert salinity into resistivity. If the water contains large amounts of other chemicals, the amounts of the various electrolytes present must be converted to their equivalent weights of NaCl. The resulting equivalent concentration is expressed in parts per million or grains per gallon of NaCl.

2. The resistivity of an electrolyte decreases as its temperature increases. This is of great importance in electrical logging because the temperature in the earth increases with depth. It is frequently necessary to compare the resistivity of a drilling mud measured at surface temperature to that of a formation measured at a much higher temperature in a deep borehole. The two values of resistivity can be compared only after they have been converted to values that would have been observed at a common temperature.

This conversion can be made by using the chart in Fig. 5. For example, if the drilling mud had a resistivity of 1 ohm-m at 60° F, the chart gives its corrected resistivity as 0.5 ohm-m at 120° F.

It is common practice in wireline logging to measure the bottomhole temperature (BHT) with a maximum thermometer contained in the body of the sonde. (The value read on the thermometer gives the maximum temperature of the mud, which is generally at the bottom of the hole.)

In case a thermometer is not available, the BHT can be approximated by using one of the two following equations. These were established empirically from the average of numerous measurements made in various geological areas.

For coastal belts, such as Louisiana, Texas Gulf Coast, and California,

$$T = 80° + 1.2 \frac{D}{100}, \qquad (2\text{-}1)$$

where

$T$ = temperature in degrees Fahrenheit

and

$D$ = depth in feet.

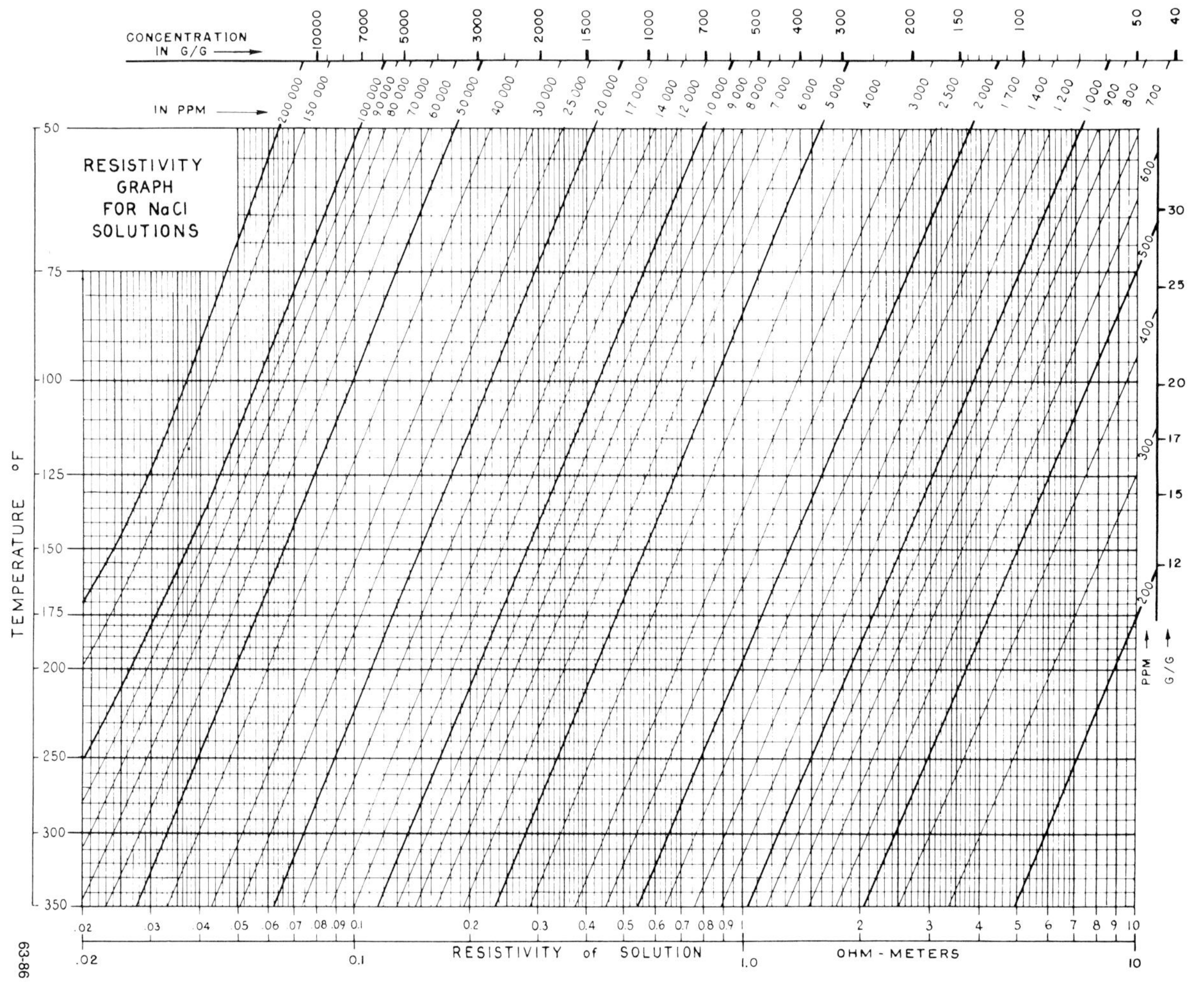


Fig. 5 Resistivity graph for NaCl solutions.

For the inland area, such as midcontinent, Rocky Mountains, west Texas, Mississippi, and Illinois,

$$T = 80° + 0.8 \frac{D}{100} . \tag{2-2}$$

**Relation of Formation Resistivity To Lithology**

The ability of an earth formation to conduct current is directly affected by the amount of water in the pores; i.e., by the porosity of the formation. If the voids inside the formation were parallel cylindrical channels, the resistivity of the formation would be inversely proportional to the porosity. Because of the relative sizes and shapes of the mineral grains and the presence of cementing material, an electrical current crossing the formation has to flow along multiple tortuous paths. The cross sections of these paths vary rapidly from comparative large values in the pores to very small values in the intervals linking the pores. The formation resistivity is, therefore, a more complex function than in the case of parallel channels. Offhand, it would seem that this function would involve not only porosity but also other characteristic parameters of lithology, such as permeability, tortuosity, and surface areas of grains. Actually, the many experiments made on various porous formations have shown that in the case of permeable formations that do not contain shaly material, the formation resistivity can be related solely to the porosity through a comparatively simple empirical equation.

In order to express this relation, it is convenient to use a parameter called "formation - resistivity factor," which is equal to the ratio $R_0 / R_w$. $R_0$ is the resistivity of a given formation entirely filled with water, and $R_w$ is the resistivity of the water in its pores. For all formations that do not contain shaly material, the formation factor is a constant regardless of the resistivity of the water.

Experiments have shown that the formation factor can be expressed with reasonable accuracy in terms of porosity by the equation

$$F = \frac{a}{\phi^m} , \tag{2-3}$$

where

$F$ = formation factor,

$\phi$ = porosity,

$m$ = an exponent, called cementation factor, which varies with the lithology,

and

$a$ = tortuosity constant.

Several values have been proposed for $a$ and $m$, but the resulting curves showing $F$ versus $\phi$ do not differ much.

In the construction of numerous graphs for log interpretation, the equation proposed by W.O. Winsauer has been used:

$$F = \frac{0.62}{\phi^{2.15}} \tag{2-4}$$

For unconsolidated formations, such as clean sands, a good approximation is

$$F = \frac{0.75}{\phi^2} . \tag{2-5}$$

For consolidated formations, such as hard sandstones and limestones, it seems preferable to take

$$F = \frac{1}{\phi^2} . \tag{2-6}$$

These equations have the advantage of simplicity in the computations (Fig. 6).

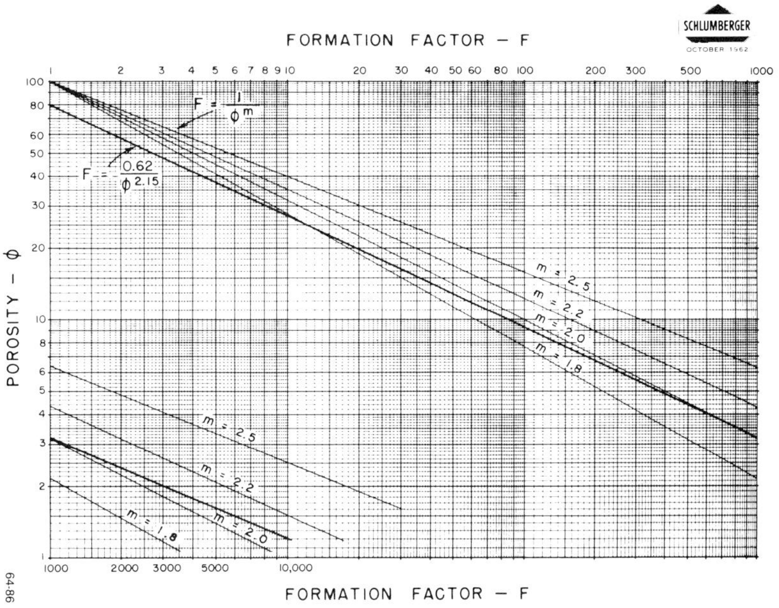


Fig. 6 Porosity-F relationship.

Limestones often contain many fissures and vugs caused by tectonic stress and leaching by circulating waters. In these limestones, it is often advisable to use relationships determined by local observations instead of the above expression.

Shales and clays are porous and usually contain mineralized water. They have, therefore, an appreciable conductivity. On the other hand, the pores are so small that practically no movement of fluid is possible. Accordingly, the shales, whether deposited in thin laminations or dispersed in the interstices of the sand grains, contribute to the conductivity of the formation without contributing to its effective porosity. The relation between formation resistivity and porosity becomes more complex than that for clean formations.

Because of the additional conductance resulting from the presence of interstitial shale, the ratio of formation resistivity to water resistivity (i.e., the formation factor) changes when the resistivity of the water changes. Nevertheless, experiments have shown that if the shale content is not too great, this ratio is almost constant for low values of water resistivities. Accordingly, the formation factor measured when the pores contain highly mineralized water can be considered, at least to a first approximation, as a characteristic parameter of the formation. It can be related to the effective porosity of clean sands through Eq. 2-5.

**Relation of Formation Resistivity to Fluid Saturation**

When a part of the pore space is occupied by an insulating material, such as hydrocarbon, the resistivity of the rock increases with respect to the value it would have if it were 100% water bearing. The resistivity of such a rock is a function of the relative proportion of hydrocarbons and connate water in the pores. The parameter generally used in well log interpretation is water saturation, $S_w$, which is the fraction of pore volume occupied by water.

The water saturation in a reservoir rock depends on many factors:

- Characteristics of the rock (porosity, permeability, surface areas of grain, etc.)
- Characteristics of the fluids (viscosity, density, etc.)
- Height above the water table of the interval under study

The water saturation generally decreases with increasing elevation in a reservoir from 100% at the level of the water table to a lower value at the top of the reservoir. This lower value is never reduced to zero because there is always a continuous network of capillary water in the pores. In other words, the resistivity of a permeable rock may reach an extremely high value, but this value is never infinite.

For primarily clean formations, the relation between formation resistivity and fluid saturation can be expressed by the following equation (Fig. 7):

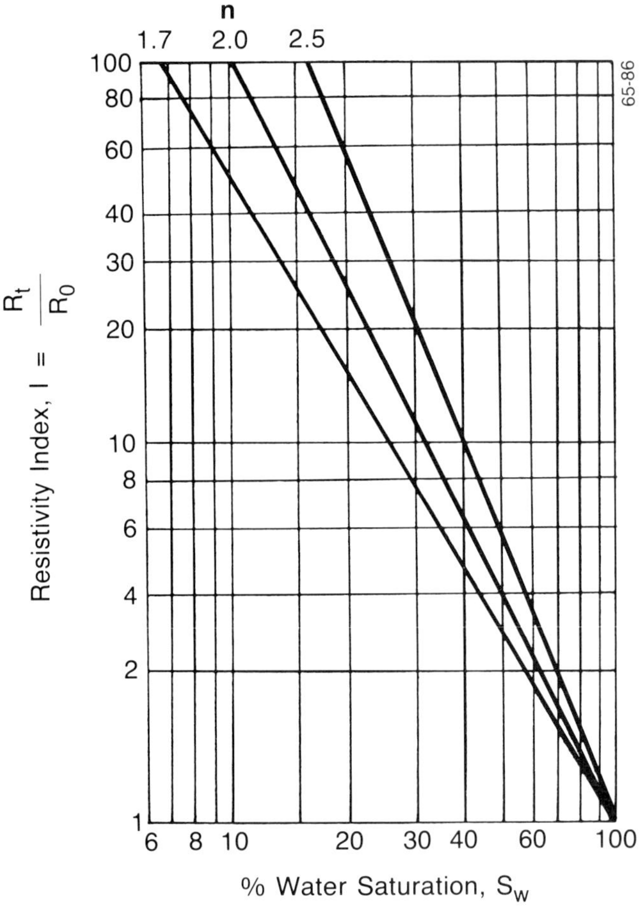


Fig. 7 Resistivity ratios.

$$S_w = (R_0/R_t)^{1/n}, \qquad (2\text{-}7)$$

where

$S_w$ = water saturation (proportion of pore space occupied by water),

$R_t$ = resistivity of the formation (containing hydrocarbons and water, with water saturation $S_w$),

and

$R_0$ = resistivity of the same formation when entirely saturated with the same water ($S_w$ = 1).

The ratio $R_t/R_0$ is sometimes designated as the "resistivity index," I. Accordingly, $S_w = I^{1/n}$.

Furthermore, since $F = R_0/R_w$,

where

$F$ = formation factor

and

$R_w$ = water resistivity,

then Eq. 2-7 can also be written:

$$S_w = (R_0/R_t)^{1/n} = \left(\frac{F\,R_w}{R_t}\right)^{1/n} \qquad (2\text{-}8)$$

The exponent $n$ varies between 1.6 and 4 depending on the type of formation. Experience shows that taking $n = 2$ gives a sufficient approximation for all practical purposes.

Eq. 2-8 would, therefore, read:

$$S_w = \left(\frac{R_0}{R_t}\right)^{1/2} = \left(\frac{F\,R_w}{R_t}\right)^{1/2} \qquad (2\text{-}9)$$

Again, the relation between formation resistivity and water saturation is more complex in shaly sands because of the additional conductance resulting from the shale network.

## PERMEABLE BED INVADED BY MUD FILTRATE

Usually, the hydrostatic pressure of the mud is greater than the natural pressure of the formations. Under these conditions, the mud filtrate invades the permeable beds.

Fig. 8A represents a schematic cross section of a permeable bed penetrated by the borehole. For the sake of simplicity, the formation is homogeneous, isotropic, and free of shaly material.

The diagram of radial distribution of resistivities is shown on Fig. 8B, where the distances from the axis of the hole are in abscissae and the resistivities are in ordinates.

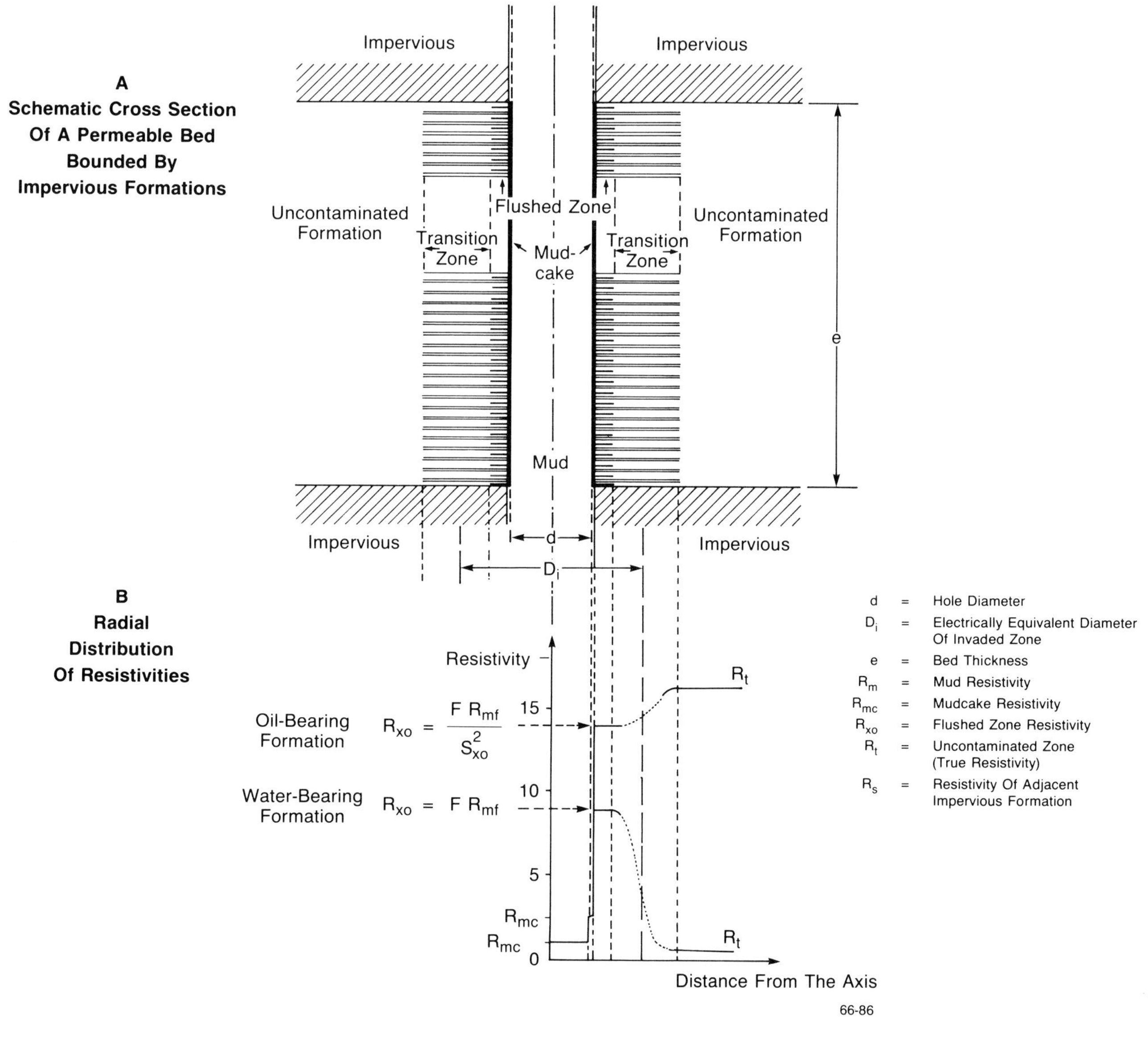


Fig. 8 Resistivity profiles.

Starting from the axis, the first two media encountered are: drilling mud, resistivity $R_m$, and mudcake, resistivity $R_{mc}$.

The thickness and the nature of the mudcake are the result of the mud and the drilling conditions, not of the formations. The thickness usually varies between ⅛ in. and 1 in. In water-based mud, the resistivity of the mudcake is usually equal to one or two times the resistivity of the mud, but it tends to be greater in weighted muds. The difference may be appreciably greater in some oil-emulsion muds.

Behind and close to the wall of the borehole, the interstitial water that originally existed in the pores has been flushed by the mud filtrate. This region, which generally extends about 3 in. from the wall, is called the flushed zone of resistivity, $R_{xo}$.

If the bed is water bearing, the pores in the flushed zone are completely filled with mud filtrate. If the formation is principally clean, the resistivity $R_{xo}$ is equal to $F\,R_{mf}$; $F$ being the formation factor and $R_{mf}$ being the resistivity of the mud filtrate.

If the bed is oil bearing, the flushed zone contains a certain amount of residual oil. In this case, $R_{xo}$ is related to the resistivity of the filtrate, the formation factor, and the water saturation in the invaded zone, as shown by the following equation derived from Eq. 2-9:

$$S_{xo} = \left(\frac{F\,R_{mf}}{R_{xo}}\right)^{1/2} \qquad (2\text{-}10)$$

$$S_{xo} = 1 - \text{ROS},$$

where ROS = residual oil saturation.

Behind the flushed zone, the distribution of fluids in the invaded zone varies continuously until a distance is reached where the formation has not been disturbed. This variable region is called the transition zone, and the distribution of the fluids in this zone is unknown.

Accordingly, there is no real definition of the depth of the invaded zone; however, it is convenient to introduce the factor $D_i$, called the electrically equivalent diameter of invasion. It is equal to the diameter of a fictitious homogeneous invaded zone of resistivity $R_{xo}$, which would have the same effect on the measurements as the actual invaded zone. This diameter $D_i$ roughly corresponds to the diameter of a cylinder located midway in the transition zone.

In interpretation we also consider the average resistivity of the invaded zone, $R_i$, which includes the flushed zone and the transition zone. The average water saturation in the invaded zone is called $S_i$ and the average resistivity of its water $R_z$, such that

$$S_i = \left(\frac{F\,R_z}{R_i}\right)^{1/2}. \qquad (2\text{-}11)$$

As a general rule, all other conditions being the same, the greater the porosity and permeability, the smaller is the depth of invasion. When $D_i$ is expressed in terms of $d$, the hole diameter, it seldom exceeds $2d$ in high-porosity sands with usual muds but attains $5d$ or more in low-porosity formations.

In water-bearing sands with large vertical permeability the depth of invasion is extremely small, often smaller than 3 in. This circumstance (frequent in the Gulf Coast) seems to be due to the difference in specific gravities between the mud filtrate and the connate water, which permits the segregation of the two fluids.

The invasion may also be extremely shallow and often is negligible in low-waterloss mud and in some gas-bearing formations.

The resistivity of the uncontaminated zone beyond the invaded zone is called the true resistivity, $R_t$. Measuring devices do not always give the value of the true resistivity directly. Other media, such as mud column, invaded zone, and adjacent formations, may have a strong influence on the measurements. The values recorded are, therefore, apparent resistivities ($R_a$), and are generally different from the true resistivities.

The water saturation in the uncontaminated zone is given by Eq. 9, repeated below.

$$S_w = \left(\frac{F\,R_w}{R_t}\right)^{1/2}, \qquad (2\text{-}12)$$

where

$S_w$ = the water saturation in the uncontaminated zone,

$F$ = the formation factor,

$R_w$ = the resistivity of the interstitial (connate) water,

and

$R_t$ = the true resistivity of the bed under study.

On Fig. 8 the resistivity decreases outward from $R_{xo}$ to $R_t$ in the case of a water-bearing sand. This feature is general since the mud filtrate, $R_{mf}$, is usually more resistive than the connate water, $R_w$. For an oil-bearing formation, the curve increases from $R_{xo}$ to $R_t$; actually, the reverse trend ($R_{xo} > R_t$) is also frequent wherever $R_{mf}$ is considerably greater than $R_w$.

## QUANTITATIVE INTERPRETATION IN CLEAN FORMATIONS

This discussion applies only to formations that contain almost no interstitial shales. The problem of shaly formations is more complex and involves other factors that are developed in another chapter.

### Saturation

The solution of Eq. 2-12, in addition to a knowledge of $R_t$, requires a knowledge of the formation factor and of the resistivity of the connate water.

In some cases, $F$ is known from direct determinations on cores. It often happens that the order of magnitude of $F$ is inferred in a given region. This is particularly true for beds that show sufficiently constant characteristics, as do many sands.

As a general rule, an approximate value of $F$ can be obtained from the measurement of the flushed zone resistivity (Eq. 2-10) provided such a zone really exists; i.e., if the filtrate has penetrated at least 3 in. into the formation.

The combination of Eqs. 10 and 12 gives:

$$S_w = S_{xo}\left(\frac{R_{xo}\,R_w}{R_{mf}\,R_t}\right)^{1/2} \qquad (2\text{-}13)$$

In this equation, the water saturation in the flushed zone must be surmised. According to samples taken from water-based mud, the residual oil saturation varies between 15% and 25% in most cases, with an average of 20%. Therefore, $S_{xo}$ can be assumed equal to 0.8 without too great an error.

**Porosity**

The formation factor given by Eq. 2-10 can be written as

$$F = \frac{R_{xo}}{R_{mf}}\,(S_{xo})^2. \qquad (2\text{-}14)$$

Combining Eq. 2-14 with Eq. 2-5 gives the porosity in unconsolidated formations,

$$\phi = \frac{0.87}{S_{xo}}\left(\frac{R_{mf}}{R_{xo}}\right)^{1/2}, \qquad (2\text{-}15)$$

where

$S_{xo} = 1 - \text{ROS}.$

Combining Eq. 2-14 with Eq. 2-6 gives the porosity in consolidated formations:

$$\phi = \frac{1}{S_{xo}}\left(\frac{R_{mf}}{R_{xo}}\right)^{1/2} \qquad (2\text{-}16)$$

An approximate value for $R_{mf}$ can be obtained from a chart based on average statistical data or, even better, can be directly measured on actual filtrate samples.

$R_w$ can be derived from the deflections of SP curves but should preferably be determined from actual samples of connate water.

$R_{xo}$ is obtained by the microdevices: MicroLog or Microlaterolog. (Limitations are explained in the corresponding chapters.)

$R_t$ is derived from the electrical log or Laterolog curves.

Before $R_{xo}$ can be determined from the MicroLog or Microlaterolog devices, $R_{mc}$ must be known. Furthermore, to evaluate $R_t$, corrections of the apparent resistivity values are often necessary.

These corrections, depending on the conditions and devices used, involve some of or all the following factors: mud resistivity, resistivity of the flushed zone, hole diameter, bed thickness, and resistivity of adjacent formations. In other words, the factors $R_m$, $R_{mc}$, $R_s$, $e$, and $d$, although not directly related to fluid saturation and porosity, are necessary intermediate elements in the interpretation. Generally, these factors are easily determined. $R_m$, $R_{mf}$, and $R_{mc}$ can be measured. $R_s$ is given directly by any resistivity log. $e$ is determined from the location of the bed boundaries. $d$ is obtained from a caliper log.

## INTERPRETATION HINTS

Although the determinations of $R_{xo}$, $R_{mf}$, $R_w$, and $R_t$ are fundamental for the most accurate quantitative analysis, a large amount of information can be obtained from the logs even if all these factors are not known.

For example, in regions of sands and shales, such as the Gulf Coast, fairly reliable interpretations have been made for years — very often without an exact knowledge of $R_t$ and even without any knowledge of $R_{xo}$.

In the Gulf Coast, many of the formations have similar characteristics. It has been possible, since the earliest time of electrical logging, to devise empirical rules of qualitative interpretation from systematic comparisons of production tests and other pertinent data with resistivity values from conventional logs. This has been true for "thick-enough" beds and "shallow-enough" invasion, where conventional curves give values approximating $R_t$. Later, the values of formation factors and connate water resistivities were derived from geological experience, and the accuracy of the interpretation was improved.

Information from other sources, such as core analysis and production results of nearby wells, can compensate for the lack of knowledge or for an imperfect determination of some of the basic parameters. Additional information of this kind should always be used to help in the interpretation and to confirm the conclusions derived from log interpretation.

In the case of consolidated sandstones or of limestones, the appropriate logs for a detailed investigation of $R_{xo}$ and $R_t$ opposite permeable sections were not always available, which makes the analysis of individual beds very difficult. However, average values of saturation and porosity can often be derived from electrical logs if the lithologic variations within the intervals are not too great. In this approach, the average resistivity of the total invaded zone is considered instead of the resistivity of the flushed zone.

# 3 The SP Curve

## GENERAL INFORMATION

The discovery of the SP in 1931 was a major event. For the first time distinction could be made between permeable and nonpermeable beds. Today's log analyst probably pays less attention to the SP curve because permeable beds can be detected with greater accuracy using other sophisticated measurements. In prospecting with old E-logs, the SP curve is a vital part of every interpretation.

The SP log is a record of the naturally occurring potentials measured in the mud at different depths in a borehole. SP logs are recorded by measuring the differences of potential between a surface electrode at an arbitrary fixed potential and an electrode in the hole whose potential varies as it is moved along the hole. The deflections correspond to the potential differences created in the mud by the SP currents and thus make it possible to characterize the formations.

Usually, the SP log consists of a baseline, more or less straight, having excursions or peaks to the left. The baseline corresponds in most cases to shales, whereas the peaks are generally opposite permeable beds. The shape and the amplitude of the peaks may be different, according to the formation, but there is no definite relationship between the magnitude of the peaks and the values of permeability. The SP curve is essentially used to indicate porous and permeable beds and to determine formation water resistivity.

## ORIGIN OF THE SP

It is generally accepted that the SP current is generated by electromotive forces of chiefly electrochemical origin that occur at the contacts between the drilling mud, the interstitial water in the pores of the permeable beds, and the adjacent shales (Fig. 9). In the following case the permeable bed is considered as practically free from interstitial shales.

Fig. 9 shows schematically a permeable sand invaded by mud filtrate and sandwiched between two very thick shale beds. The arrows represent the various electromotive forces (emf). The direction of each arrow indicates the positive sign of the emf; i.e., the direction along which the presence of an emf corresponds to an increased voltage. This is also the direction of current circulation.

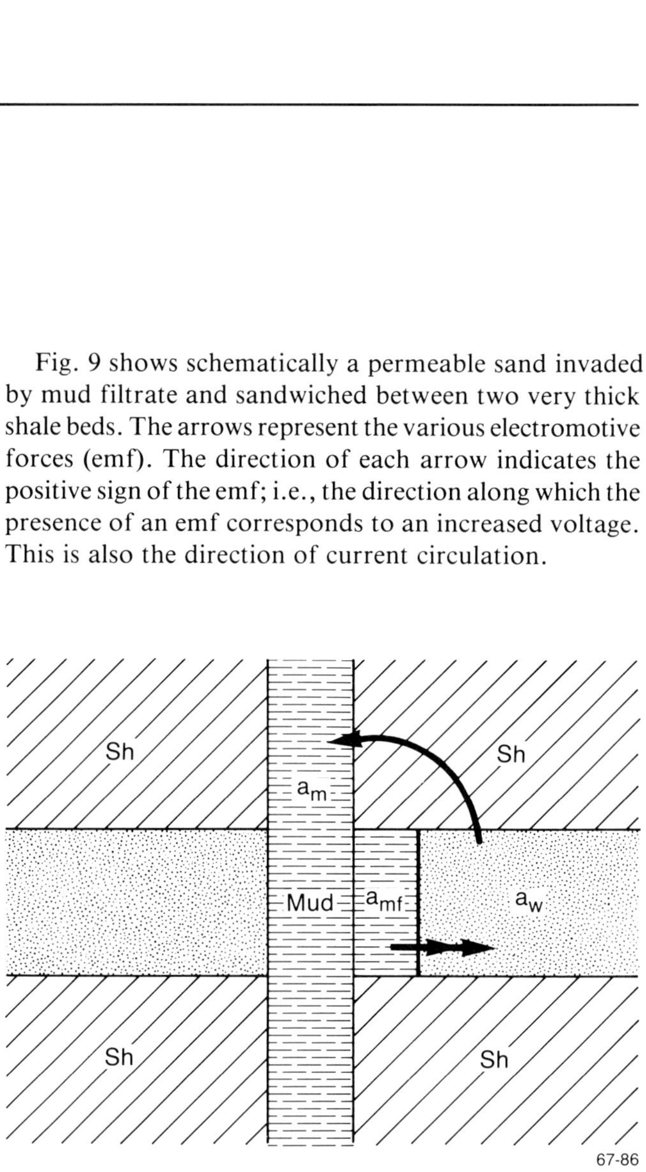


$\longrightarrow$ Membrane Pot. $= 59 \log \dfrac{a_w}{a_m}$ at 75°F

$\twoheadrightarrow$ Liq. Junct. Pot. $= 12 \log \dfrac{a_w}{a_{mf}}$ at 75°F

$a_m = a_{mf} \quad E_c = -71 \log \dfrac{a_w}{a_{mf}}$ at 75°F

Fig. 9 Electrochemical emf assuming mud and formation water are pure NaCl solutions (clean sand).

The emf between the formation water and the mud across the adjacent shale corresponds to a "membrane" potential. The emf at the contact between the mud filtrate and the formation water corresponds to a liquid junction potential. If the mud and the connate water can be considered solutions of pure NaCl, the values of these emf's at 75° F are respectively equal to $-59 \log a_w/a_m$ and $-12 \log a_w/a_{mf}$, where $a_w$ and $a_{mf}$ are the cation "activities" of the formation water and of the mud filtrate.

Generally, the activities of the mud and mud filtrate are equal so that the total emf is:

$$E_c = -71 \log \frac{a_w}{a_{mf}} \text{ at } 75° \text{ F} \quad (3\text{-}1)$$

For very diluted solutions, an equation such as Eq. 3-1 would involve the ratio of salt concentrations. The activities can be defined, for simplicity, as quantities that make such an equation correct for all concentrations. Activities, at any rate, have basic significance and can be derived from quite different sorts of measurements.

The important point is that activities are closely connected with resistivities. Therefore, the SP curve can be used to determine formation water resistivity, which is a basic factor in quantitative interpretation.

There is an additional emf, the electrokinetic emf resulting from the invasion of mud filtrate through the pores of the sand. It has long been thought that the electrokinetic, or electrofiltration, emf is practically negligible in high-salinity connate water, but that it constitutes an important part of the total emf in comparatively fresh connate waters. Experience shows that the electrokinetic emf is significant only in exceptional cases.

## CIRCULATION OF SP CURRENTS

Fig. 10 shows schematically the circulation of the SP current across the same formations as those shown on Fig. 9. Each line represents a line of current flow.

Each current line crosses the three boundaries A, B, and C. In the usual case where the salinity of the mud is higher than the salinity of the connate water, $E_c$ is positive and the current circulates in the direction of the arrows (from inside the borehole towards the permeable bed). The potential difference, therefore, between a point at the center of the bed and the adjacent shales is negative, as shown on the curve on the left of the drawing.

Along its path, the SP current must force its way through a series of resistances, both in the formation and in the mud. Along a closed line of current flow, the total of the ohmic potential drops is equal to the algebraic sum of the electromotive forces encountered.

Also, since the magnitude of the current flow is constant all along its path, the potential drop varies according to the resistance of the section through which it flows. This means that the total potential drop (which is equal to the sum of the electromotive forces) is divided between the different formations and the mud in proportion to the resistivities encountered by the current in each respective medium. The potential drop in the mud of the borehole is a measure of only part of the total emf—a great part in fresh muds—because the electrical resistance offered by the mud is much greater than that offered by the formations.

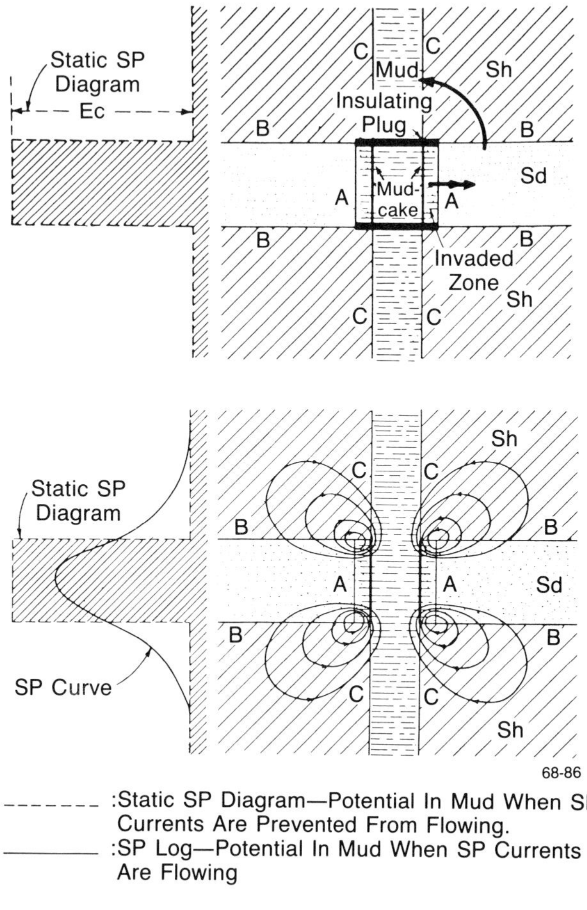


Fig. 10 Schematic representation of potential and current distribution in and around a permeable bed.

## THE STATIC SP

The value of the electromotive forces that produce the SP currents, and which therefore determine the SP log, are indicated by an idealized representation in which the SP current is prevented from flowing by means of insulating plugs, placed as shown at the top of Fig. 10. The value of the potential within each medium enclosed by boundaries or plugs would be constant. However, the potential would vary in each medium, with the difference of the potentials in the different media being equal to the emf existing at their common boundary.

If the potentials in the mud could be measured in this idealized case, a log such as the dashed, crosshatched curve on the left of Fig. 10 would be recorded. Such a diagram, plotted for the case in which no current is allowed to flow, is called a "static SP diagram." The difference of the potentials in the mud between the plugs and outside the plugs is called the "static SP" of the bed and is equal to $-E_c$.

## FACTORS INFLUENCING THE SHAPE AND AMPLITUDE OF SP PEAKS

As shown in Fig. 10, the current circulates in the mud not only opposite the permeable formation but also a short distance beyond its boundaries. As a result, although on the static SP diagram the boundaries of the permeable beds are indicated by sharp breaks, the SP curve exhibits a more progressive change in potential along the borehole and extends beyond the boundaries of that bed.

An analysis of the circulation shows that the boundaries of the layer are located at the level of the inflection points on the SP log. This fact provides a means of determining the thickness of a bed from the SP log.

Since the SP log records only that portion of the potential drop occurring in the mud, the amplitude of the SP log approaches the amplitude of the static SP diagram only when the resistance offered to the current by the bed itself and the adjacent formations is negligible compared with the resistance of the mud in the borehole. This occurs when the bed is sufficiently thick.

The shape and amplitude of the SP curve opposite a given bed may be influenced by the following factors:

- The total emf's involved (static SP)
- The thickness of the bed
- The resistivity of the bed, of the surrounding formations, and of the mud
- The diameter of the borehole
- The depth of invasion
- The presence of interstitial shale

The SP log is also influenced by a lack of homogeneity of the mud. A change in mud salinity at a certain level would result in a baseline shift at that level. However, such changes in mud salinity are seldom encountered.

With other factors remaining the same, a change of the total emf's affects the amplitude but does not otherwise modify the shape of the SP curve.

In practice, the emf's may vary from one hole to another because the salinity of the mud, and/or of the formations, is quite different. In a given hole, however, for the same type of formations and for comparable depths, there is a definite tendency for the total emf's to be the same for all beds. Freshwater sands or high-salinity sands will show, respectively, small- or large-amplitude peaks. The polarity of the peak may even reverse when the water in the sand is less salty than the mud.

Permeable beds of different porosity or with different grain size give the same emf's if the other factors remain unchanged. The emf's are also independent of permeability, even down to fractions of 1 md.

How these factors influence the SP log is explained first for the case of permeable beds situated between shales, as shown in Fig. 10. In this case the resistivities of the permeable beds are generally greater than, or about equal to, those of the shales. This case is typical of the so-called "soft formations," such as sand and shale series. The case of hard formations is discussed later.

## PERMEABLE BEDS ADJACENT TO SHALES (SOFT FORMATIONS)

Theoretical computations and field experience have shown that the amplitude of the SP deflection is almost equal to the static SP when the permeable beds are very thick and when the resistivities of the formations are not too great with respect to that of the mud. In this case, the SP curve locates bed boundaries with great accuracy. The amplitude of the deflection is less than the static SP for thin beds, and the smaller the bed thickness, the smaller the deflection.

On the other hand, when the resistivity of the formations, $R_t$, is considerably higher than that of the mud, $R_m$, the SP curves are rounded and the bed boundaries are more difficult to locate from the SP curve. With all other conditions being the same, the amplitude of the peak is less than when the ratio $R_t/R_m$ is close to one.

Fig. 11 shows theoretical data illustrating the influence of the bed thickness and the resistivity of the formations. For simplicity, in this comparison the value of the static SP is $-100$ mV for all permeable beds and the depth of invasion in the permeable beds is shallow enough to be negligible.

Fig. 11 also shows that the SP curve spreads a considerable distance outside the boundaries of the layer when $R_t/R_m$ is large, and the higher the ratio, the greater this effect.

The mud resistivity has a predominant influence on the magnitude of the SP deflection. The salinities of connate water are usually very high; so, if the mud is of high salinity the electrochemical potentials are very small. Also, for a given static SP, the lower the mud resistivity, the smaller the ohmic drop in the mud and the wider the deflection above and below the permeable beds. In fact,

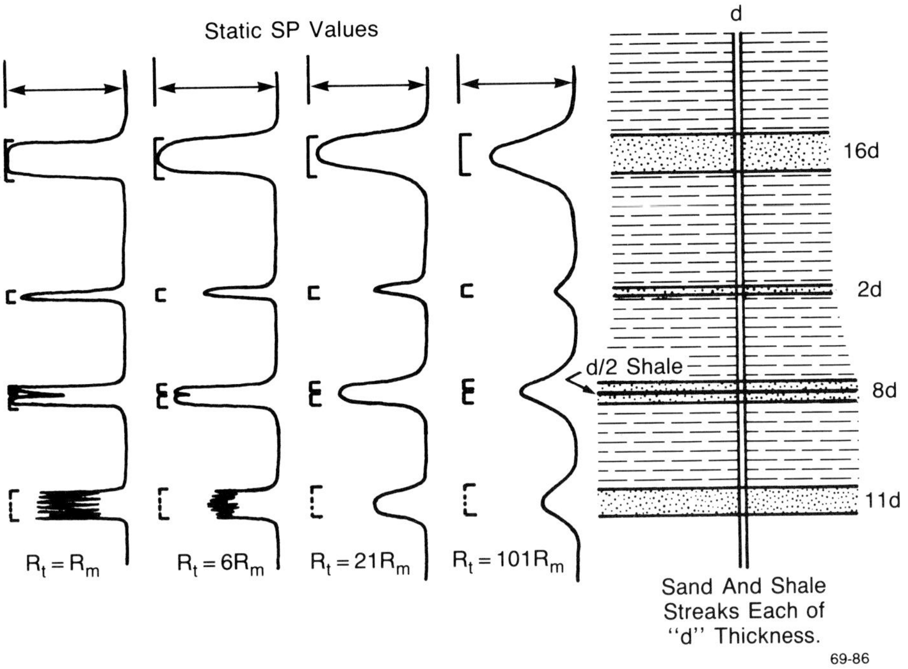


Fig. 11 Effects of bed thickness and resistivity on self-potential.

the SP curves show very small deflections and are almost flat when the boreholes are drilled with very conductive muds. In comparatively fresh formation water, conductive mud may give a reversed SP with deflections toward the positive side.

The computed curves shown on Fig. 11 are for successive beds having similar values of resistivity. In practice, this is not always the case, and the resistivities of successive beds may differ widely; for instance, when shales are interbedded with oil-bearing sands. In such a case, the SP curve is more rounded in the more resistive formation than in the other one, and the inflection point is displaced toward the bottom of the peak if the peak corresponds to the more resistive formation, and vice-versa.

An increase in hole diameter has an effect on the SP log similar to an increase in the ratio $R_t/R_m$. It tends to round the deflections and to reduce the amplitude of the peaks opposite the beds.

In permeable beds, the emf's originate inside the formation at the boundary between the mud filtrate and the formation water. As a result, deep penetration of mud filtrate has an effect on the SP log similar to an increase in hole diameter. The SP peaks are wider than they would be in the case of shallow invasion, and the amplitude of peaks corresponding to thin permeable beds is smaller than for the case of shallow invasion.

## HARD FORMATIONS

The case of permeable beds in compact and highly resistive formations, such as limestone, is of particular interest.

The SP log in hard formations has a shape difficult to understand at first glance. These formations are highly resistive except for conductive zones of two types: the permeable zones, whether oil bearing or water bearing, which are more conductive because of the high-salinity capillary water in their pore spaces; and shales.

Fig. 12 represents schematically a sequence of permeable beds, compact beds, and shales.

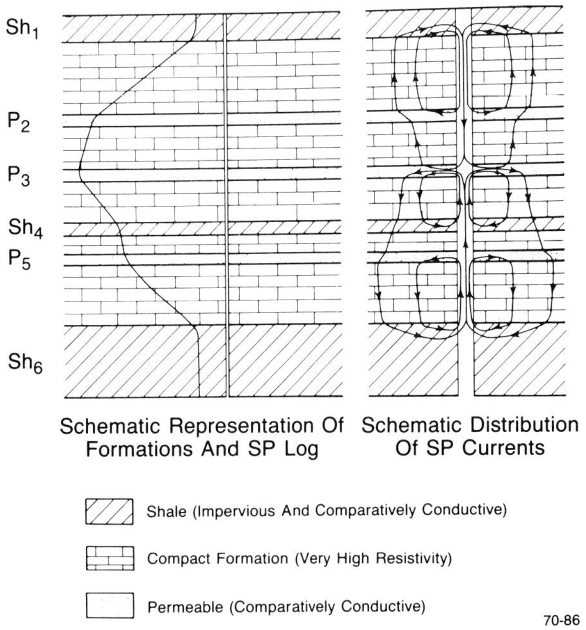


Fig. 12 SP development.

The circulation of the SP currents is also represented schematically. The SP currents, which are generated by the different emf's, flow into the permeable sections. They cannot flow through the adjacent hard formations in sections close to the borehole because of the large resistances. The SP current penetrates more deeply in the permeable beds and enters the hard formations through larger cross sections. The currents continue their path in the hard formations to the shale beds, through which they return to the borehole, and then pass through the mud back to the permeable beds to complete their circuits. They cannot come back to the mud through other permeable beds because they would encounter emf's that would oppose the flow of current in that direction. When the first conductive beds they encounter are permeable, they simply cross them until they reach the shale beds. This is the case for the SP currents that penetrate the permeable bed $P_3$; those flowing downward come back to the borehole partly through the shale $Sh_4$; some cross the permeable bed $P_5$ in order to reach the shale $Sh_6$. It thus appears that the peaks corresponding to the permeable zones spread above and below these zones in an abnormal manner, and it is impossible to accurately determine the boundaries of the permeable zones.

Furthermore, all along the borehole opposite a hard formation, the current in the mud column remains essentially constant, as does the drop of potential per unit length of hole. This gives a constant slope, as shown by the straight lines of these portions of the SP on Fig. 12. At the level of each conductive bed some SP current generally enters or leaves the hole — and the slope of the SP log is modified. The SP log changes its slope at the level of the permeable bed $P_2$ because part of the current leaves the hole and flows into that bed.

As a general rule, permeable beds are characterized on the SP log by slope changes with the convexity toward the negative side of the log. Shale beds are characterized by slope changes with the convexity towards the positive side of the log. Highly resistive formations correspond to fairly straight parts of the SP curve.

## QUANTITATIVE INTERPRETATION

The first step in a quantitative interpretation is to determine the static SP at the level of the bed under consideration. The preceding material has shown that the deflection of the curve gives the value of the static SP only where the bed is sufficiently thick. When contrasts in resistivities between the bed and the adjacent formations and between the formations and the mud are comparatively low, as is frequently the case in sand-shale series, the required thickness is about 5 to 6 ft.

In consolidated or hard formations, the required bed thickness may be much greater. However, this thickness includes the hard rocks that may separate a permeable section from the adjacent shales. Even if the permeable section is very thin, the static SP can be obtained if the overall thickness between the bounding shales is great enough.

When the static SP is known, which is the case when the bed is sufficiently thick, the determination of formation water resistivity is based on the equation

$$SSP = -K \log \frac{a_w}{a_{mf}}, \qquad (3\text{-}2)$$

This, as already said, is true provided the mud and the formation water can be considered pure NaCl solutions. Experience shows that this condition is fulfilled for all muds except those that contain a significant proportion of calcium (such as gypsum-based mud).

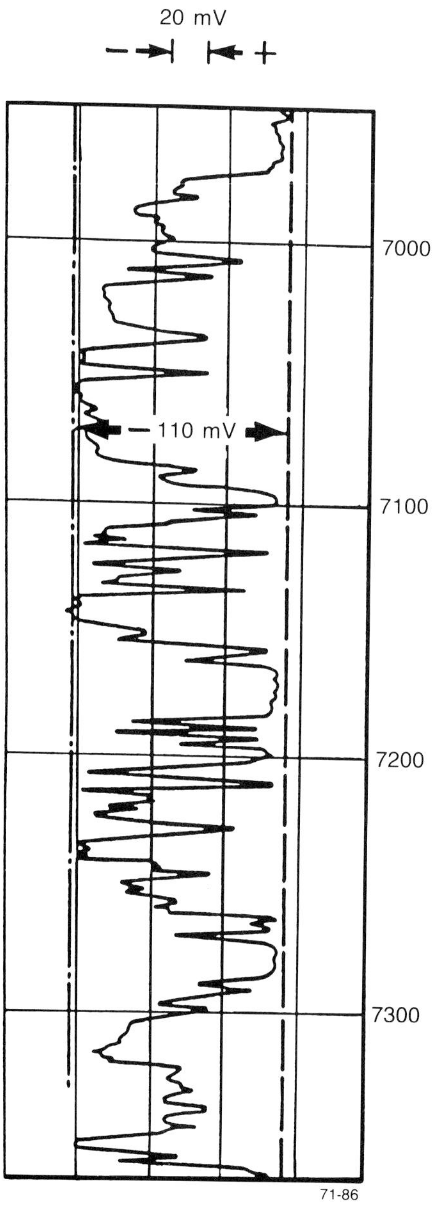


Fig. 13 Typical SP curve.

The formation waters, in most cases, actually contain certain salts different from NaCl. Laboratory data and field experience have shown that the effect of these salts is practically negligible when the water resistivity is less than about 0.3 ohms at 75° F.

$R_w$ is determined from the SP curve by using Charts A-10 and A-12 in Appendix A. A work example illustrates the procedure: Refer to Fig. 13, which is a typical SP curve. The static SP is the maximum deflection observed over the interval of interest. In this case, it is 110 mV. The measured BHT is 150° F at 7100 ft and the measured $R_{mf}$ is 0.72 ohm-m at BHT (from the log heading).

Chart A-10 is a graphical representation of Eq. 3-2 with the static SP plotted in abscissa on a linear scale and $R_{mf}/(R_w)_e$ in ordinate on a log scale. Several curves are traced that correspond to different temperatures. Enter the value of the SP (110 mV); proceed upward to a line equivalent to the formation temperature (150° F). On the left margin read $R_{mf}/(R_w)_e$. This value is 24. $(R_w)_e$ then $= R_{mf}/24 = 0.72/24 = 0.030$. Chart A-12 will convert $(R_w)_e$ to $R_w$. Enter $(R_w)_e = (0.030)$ in the abscissa and proceed upward to an equivalent temperature (150° F) and read $R_w$ at the scale on the left margin. In this case $R_w = 0.041$.

The accuracy of the technique depends, of course, on the accuracy of the input values. There should be few problems with temperature and $R_{mf}$ values; the major source of error is in finding an appropriate static SP value. If in doubt, compare the answer with cataloged water samples, which are usually the most accurate source of $R_w$. Additional examples of $R_w$ determination are given in the charts in Appendix A.

## SUMMARY - THE SP CURVE

**Measures:**

- The naturally occurring potentials in the well that are developed opposite permeable beds.

**Features and Applications:**

- Can detect permeable beds.
- An excellent correlation curve.
- Used to determine values of formation water resistivity, $R_w$.
- A good shale indicator.

**Limitations:**

- Not developed in empty holes or oil-based mud systems.
- Poorly developed in salty muds.
- Sensitive to equipment noise sources.

# 4 Conventional Resistivity Methods

Electrical logs, as well as the other methods using electrodes, were recorded in holes drilled with water-based drilling mud that provided the electrical contact necessary between electrodes and formation.

The relative positions of the electrodes making up the sonde have a bearing on its general response. Depending on the distribution of electrodes, a device may, for instance, be good for evaluating thin beds but may not give a good evaluation in the thick layers. Another device that provides a better value of $R_t$ may not be suitable for locating formation boundaries.

In the early days of electrical logging, one single resistivity device was used. Later, it was found best to employ three different electrode arrangements together so that the resistivity curves recorded gave as complete a picture as possible of all formations encountered in the borehole, however different their characteristics.

Three types of measuring devices, referred to as "conventional," were used on electrical logs: the normal device, the lateral device, and, in some hard-rock territories, the limestone sonde.

As shown in Fig. 14, the normal device consists of a small current electrode (A) and a small measuring electrode (M) separated by a short distance (AM) called the spacing of the device. Current is passed between A and an electrode (B) located at the surface; the potential difference created by the flow of current is measured between M and an electrode (N) that is remote from both current electrodes. The effect of N on the measurements is negligible except in thick, extremely resistive formations. Therefore, potential difference measured is practically equal to the potential of M.

Fig. 15 shows the lateral device; the potential difference between M and N, caused by the flow of current from A, is measured. Since the distance separating M and N is small in comparison with the distance between A and the midpoint of MN (O), the quantity measured is little different from the electrical field strength, or the opposite of potential gradient, at Point O. The distance (AO) is called the spacing of the sonde.

For convenience, in actual field equipment the equivalent electrode arrangement is used for recording lateral curves. In this arrangement, the positions of the current electrodes and measuring electrodes have been interchanged and are termed the "reciprocal arrangement."

In most locations, the three resistivity curves of the conventional electrical log were recorded by a short normal device (AM = 16-18 in.); a long normal (AM = 64 in.)

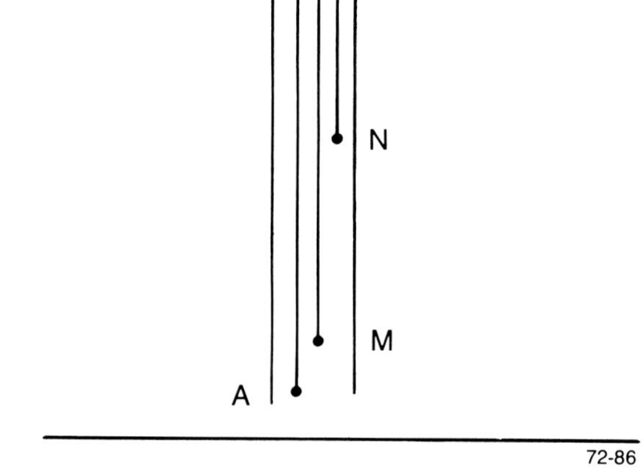


Fig. 14 Schematic of normal device.

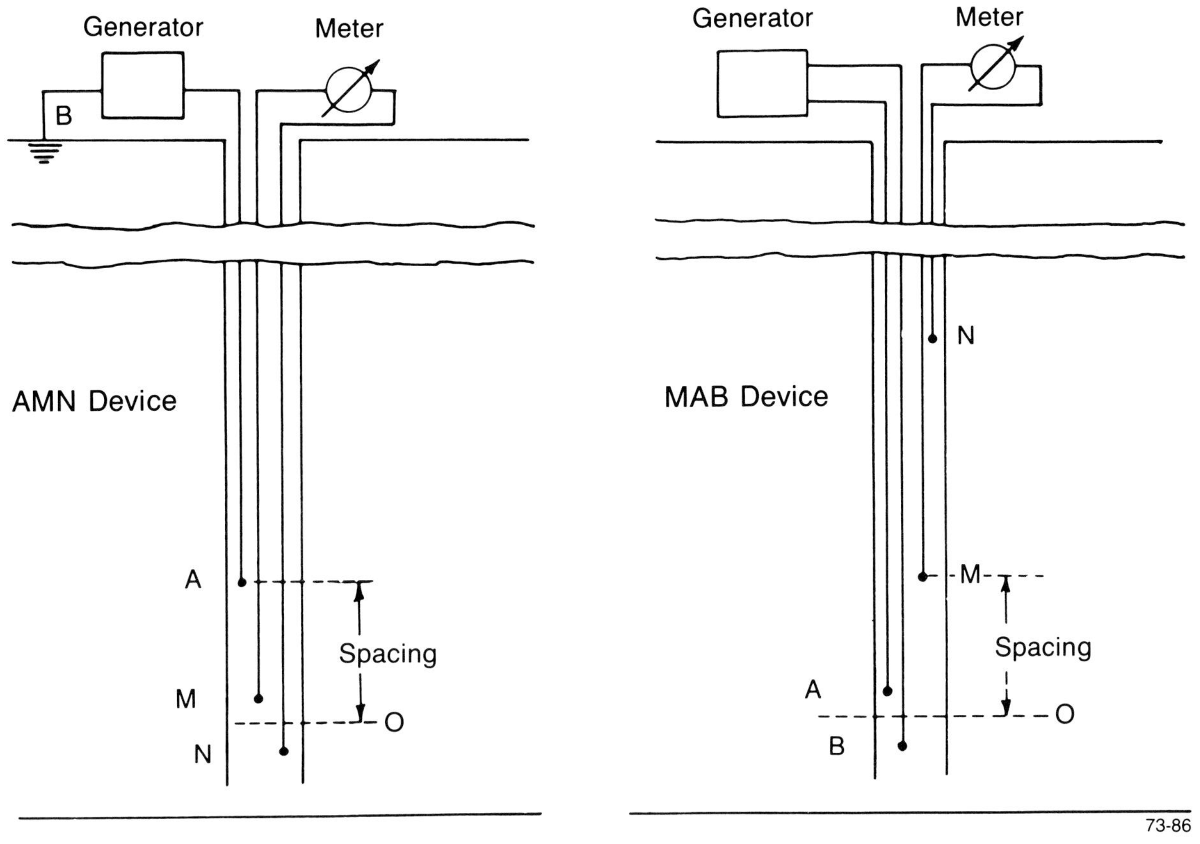


Fig. 15 Schematic of lateral device—theoretical and actual.

or, in the Pacific Coast area, a short lateral (AO = 6-9 ft); and a long lateral (AO = 18 ft 8 in.). In some regions, such as the Permian basin (west Texas and New Mexico), the short normal was reduced to 10 in. and the limestone sonde was used instead of the long normal. Also, in some areas a 24-ft lateral was used. The log heading identifies the log type and spacing, but they can usually be determined by observing the curve responses.

A knowledge of how the measuring devices respond under various conditions (thick beds, thin beds, invasion...) is necessary for a correct interpretation of the logs.

First, the characteristic shapes of the curves should be known for discrimination of beds, definition of boundaries, and correlation. Second, in quantitative analysis, the true resistivity is a basic factor seldom given directly — so it is important to know what relationships exist between the readings and the true resistivity.

The response of a measuring device can be treated mathematically only by using several simplifying conditions:

1. The beds are homogeneous and bounded by parallel planes that are perpendicular to the axis of the sonde: this amounts to neglecting borehole effect and invasion.

2. The beds are "infinitely thick" and the invaded zone is homogeneous and limited by a cylinder coaxial with the hole.

The response in beds of finite thickness, invaded or not, cannot be solved exactly by computations. Departure curves were computed with some approximations for a limited number of cases. A more complete solution was sought in early times through laboratory models using different media with electrolytic conductivities. Later, resistor network systems were used to help resolve the problem.

## SHAPES OF NORMAL AND LATERAL CURVES

Fig. 16 shows the curve recorded in the laboratory with normal devices for homogeneous resistive layers sandwiched between beds of low resistivity. The point of measurement is midway between A and M on the sonde.

It can be seen immediately that the curves are symmetrical with respect to the center planes of the layers. This is a general feature of the normal device. As a matter of fact, the same curves are recorded if M is above A instead of A above M, as indicated in the figure.

The upper part of the figure illustrates a bed thicker than the spacing (bed thickness $e = 6d$, spacing AM $= 2d$, $d$ = hole diameter). The bed boundaries are not sharply indicated on the resistivity log; they are rounded from the influence of the borehole. Also, the thickness of the bed, as indicated by the distance between the two points of inflection (P and P′) on the curve, is less than its actual thickness by an amount equal to the spacing. The error in picking the boundaries of thick resistive beds is small for normal curves of short spacings, and this was one reason for recording a short normal. Remember, however, that normal curves tend to show resistive beds thinner than they actually are by an amount equal to the spacing. In a similar manner, they tend to show conductive beds thicker than they actually are by an amount equal to the spacing.

The lower part of the figure shows a resistive layer thinner than the spacing. A depression is opposite the layer and two symmetrical small peaks (c and d) are on each side of the depression. This feature illustrates the main disadvantage of the normal device: beds thinner than the spacing, no matter how resistive they may be, appear conductive on the logs.

Fig. 17 shows the corresponding curves for lateral devices. The point of measurement is O, midway between the electrodes (M and N).

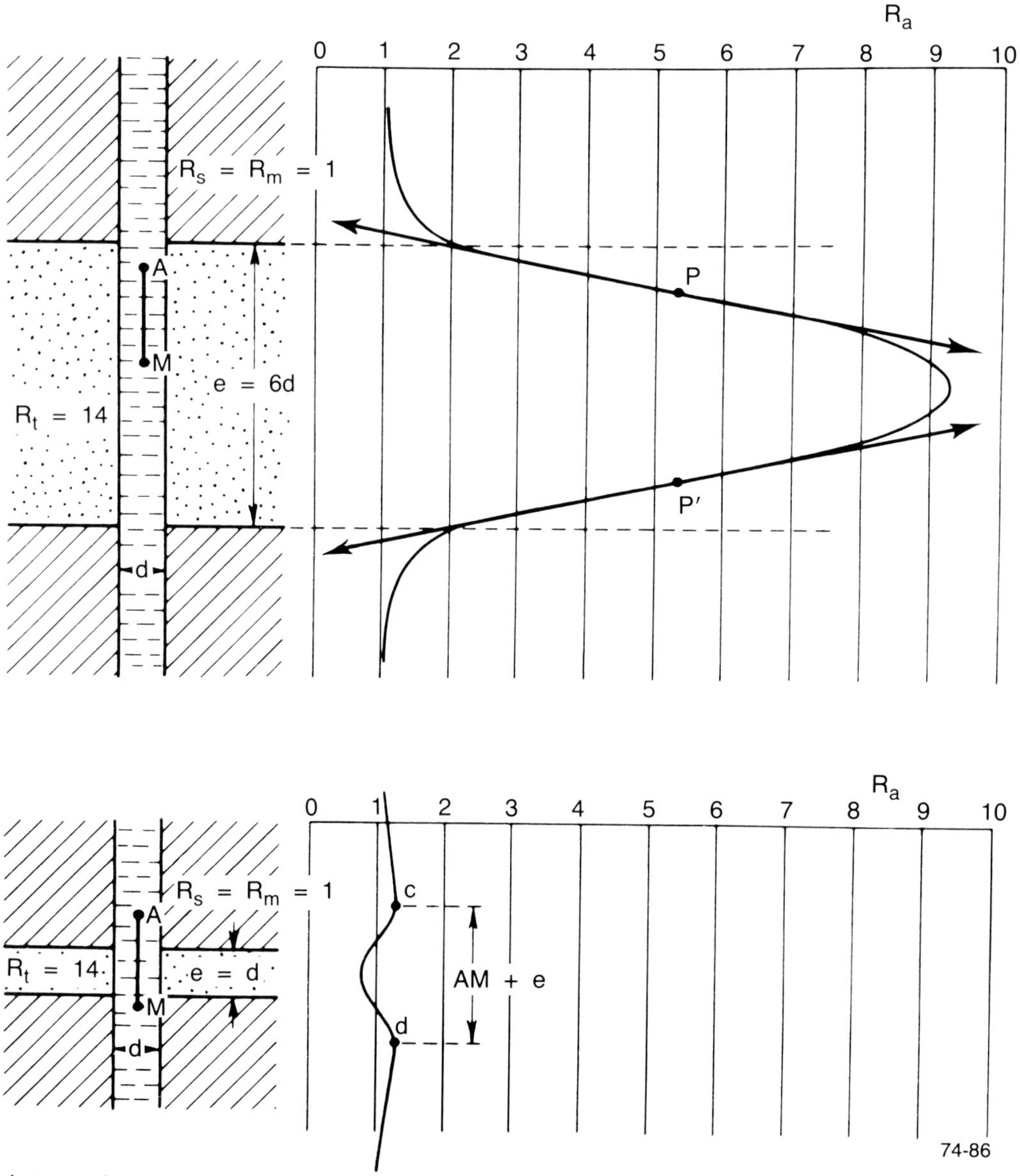


Fig. 16 Normal curve shapes.

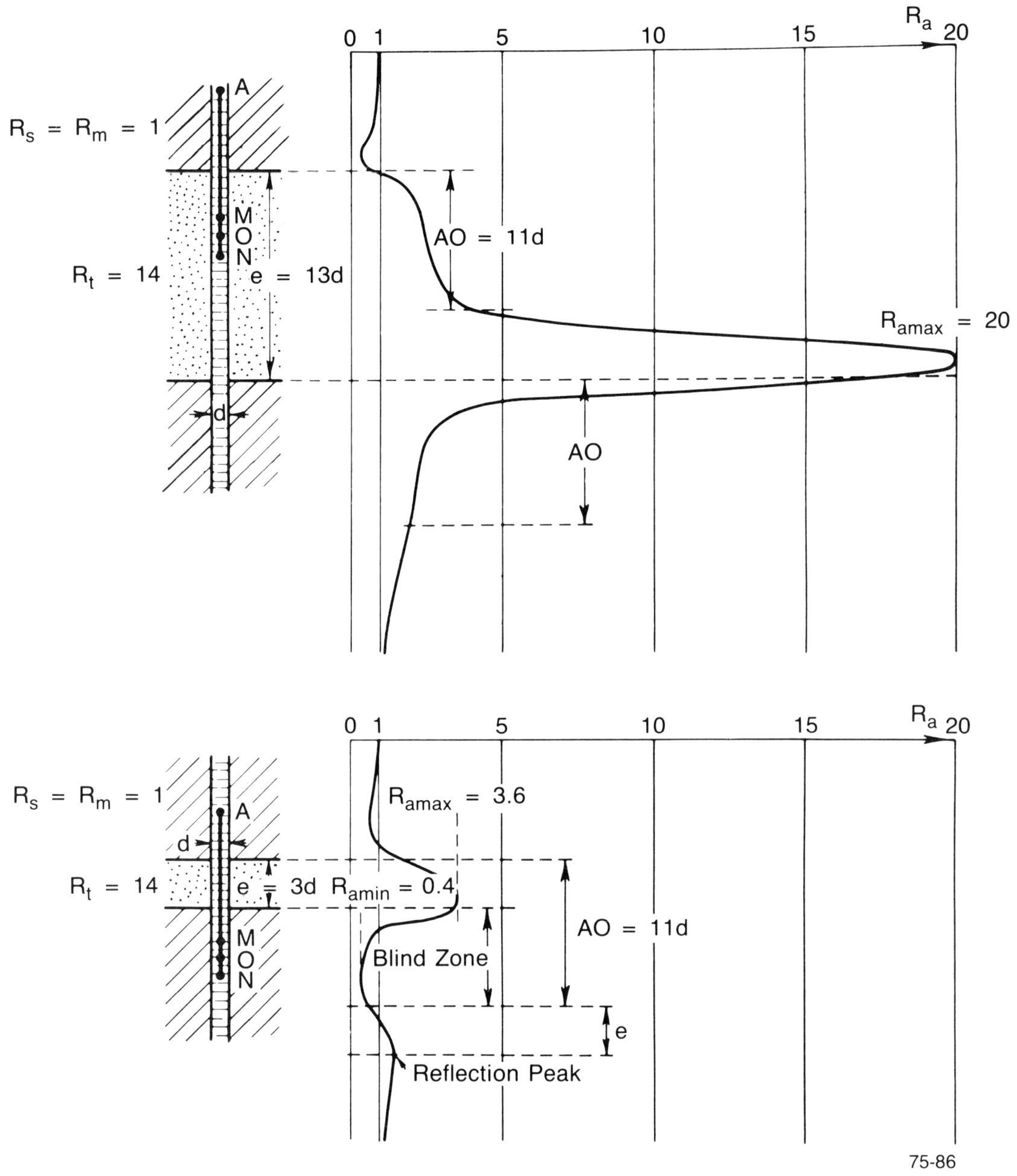


Fig. 17 Lateral curve shapes in resistive beds.

In contrast to curves recorded with normal sondes, the lateral curves are markedly dissymmetrical with respect to the center planes of the layers, and their features are considerably more complex. As before, the transitions in the curves corresponding to formation boundaries have been rounded by the borehole effect.

For the case of a bed thicker than the spacing, the upper boundary of the bed is not well defined by the lateral curve and, as a whole, the bed appears to be displaced downwards. The amplitude of the shift is approximately equal to the spacing.

The lower part of Fig. 17 corresponds to a resistive layer thinner than the spacing. The bed is indicated by a sharp peak of comparatively low apparent resistivity. There is a slight depression above the layer, but a more striking feature is the zone of low resistivity below the layer, followed by a second smaller peak located below the bottom boundary of the layer at a distance equal to the spacing. This secondary peak is called a "reflection peak," and the zone of a very low apparent resistivity is called the "blind zone." The blind zone corresponds to the interval during which the resistive streak is located between the current electrode and the measuring electrodes.

The lateral is useful for locating thin resistive streaks. The interpretation, however, may be difficult in a sequence of several resistive streaks close to each other; some beds can be missed and reflection peaks can be mistaken for actual resistive streaks in the formation.

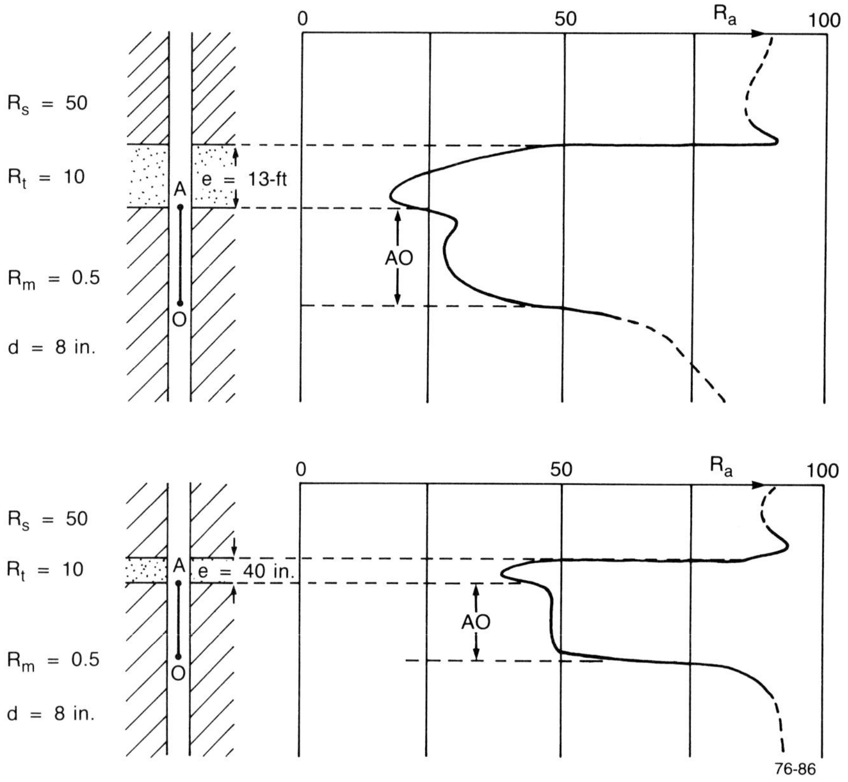


Fig. 18 Lateral curve shapes in conductive beds.

Fig. 18 shows the behavior of a lateral device in beds less resistive than the surrounding formations. The curves were determined by using a resistor network.

Fig. 19 is a schematic representation of the behavior of the normal and lateral devices opposite a thick, highly resistive formation, such as a limestone, containing conductive zones.

To understand the behavior of these curves, remember that opposite a highly resistive formation a great part of the current from A flows along the borehole, dividing in inverse proportion to the resistance of the mud column up the hole and down the hole. The resistance of each path is determined mostly by the resistance of the mud column in the hole between the current electrode and the nearest conductive bed. When the current reaches the conductive bed, it has, depending on the thickness and the conductivity of the bed, a relatively low resistance path away from the hole.

This circumstance can be used to explain why the lateral device, which has Electrodes M and N below the current electrode, records a curve that is lopsided in a downward direction. The deeper the device is into the resistive bed the greater is the proportion of the current flowing downward in the hole from A, and therefore the greater is the potential drop in the mud measured between M and N. The apparent resistivity corresponding to this potential drop will thus be largely dependent on the size of the hole, the resistivity of the mud, and the position of the device with respect to the boundaries of the highly resistive zones.

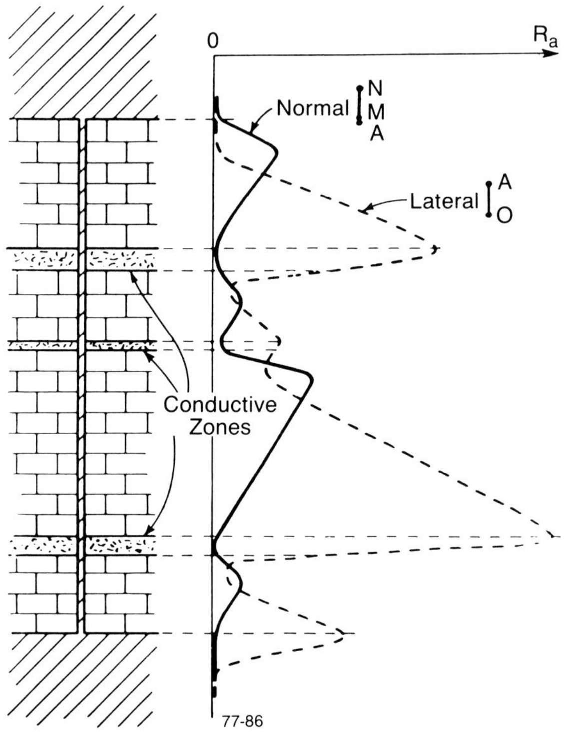


Fig. 19 Curve shapes in hard formations.

The unsymmetrical appearance of the normal curve is explained in a similar manner for the three-electrode arrangement illustrated, in which M and N are above the current electrode. As soon as all three electrodes are inside the hard formation, the upper electrode is no longer near zero potential, as is basically assumed for the normal, and the arrangement actually measures the potential drop caused by the current flowing in the mud between M and N.

Therefore, three-electrode devices give unsymmetrical curves in hard formations, and the direction of the lopsidedness depends on whether the measuring electrodes are above or below the current electrodes. Along with this dissymmetry, Fig. 19 shows that for the lateral, the positions of the minima are shifted quite noticeably downward with respect to the exact depths of the conductive layers, and for the normal a slight upward shift of the minima is also discernible. This often results in the maximum of one curve matching in depth the minimum of the other curve. Also, the depressions on the normal curve opposite the conductive layers are very smooth and considerably broader than the thickness of the beds, and the boundaries of the limestone formations are poorly marked.

## THEORETICAL BASES OF THE QUANTITATIVE ANALYSIS

Coming back to the basic formula that gives the water saturation in terms of the electrical factors given by the logs,

$$S_w = \left(\frac{F\,R_w}{R_t}\right)^{1/2} = S_{xo}\left(\frac{R_{xo}\,R_w}{R_{mf}\,R_t}\right)^{1/2},$$

one essential factor to be obtained is $R_t$, true resistivity of the uninvaded part of the bed.

Laboratory results shown in preceding figures indicated that resistivity measured by a given device is dependent on the geometry and the resistivities of the media near the electrodes and may thus differ considerably from the true resistivity of the formation. This measured resistivity value is called "apparent resistivity" and is designated by the symbol $R_a$.

## ESSENTIAL RESPONSES OF THE MEASURING DEVICES

All these data, obtained by computations or through laboratory experiments, are extremely valuable for evaluating the performances of measuring devices under the various conditions encountered in field practice.

For example, the following values were used for the parameters: $R_m$ = 1 ohm-m; $R_{xo}$ = 100 ohm-m; $R_t$ from 5 to 200 ohm-m for water saturations from 100% to 15% (average conditions in many hard-rock territories); AO = 18 ft 8 in.; $d$ = 8 in.; $D_i$ = 80 in. If the formation is very thick, the departure curves will show that the lateral device would give apparent resistivities ranging from 40 to 300 ohm-m. In other words, the lateral device will give readings not too far from $R_t$ in oil-bearing formations, but much too high in water-bearing formations. It could be concluded that the lateral device would lead to a too optimistic interpretation of saturation in this case.

Suppose the mud is more conductive; for example, $R_m$ = 0.1 ohm-m. This would correspond to a value of about 10 ohm-m for $R_{xo}$, with all other conditions being the same. The departure curves show that the lateral device would read, respectively, 10 and 150 compared to 5 and 200 ohm-m. The apparent resistivities in this case are much closer to the true values.

Highly resistive muds are not favorable for recording conventional electric logs. Highly conductive muds are not desirable either. In these muds the normal and the SP curves are reduced to almost flat lines; the departure curves and field experience show that the optimum mud resistivity is equal to about five times that of the formation water resistivity.

Some essential conclusions (confirmed by field experience) can be derived from these theoretical and experimental data.

1. The ability of the measuring devices to determine $R_t$ is limited primarily by bed thickness and depth of invasion. The limitations are more severe when the contrast is great between the resistivity of the bed and the resistivity of the adjacent formations, and between the resistivity of the formations and that of the mud column. This is why electrical logs are more efficient in soft sand-shale series (resistivity contrasts generally low, invasions shallow) than in hard-rock territories where contrasts are high and invasion is usually deep. In sand-shale series, even in so-called soft formations, the sands are often adjacent to or interbedded with thin resistive streaks, such as consolidated sandstones or lignites, that introduce appreciable distortions into the resistivity curves.

2. All in all, the long spacings necessary to overcome deep invasions are adversely affected by the adjacent formations. The short spacings are less affected by bed thickness but are sensitive to invasion. Accordingly, the short normal is good for delineating beds but is not always suitable for investigating true resistivities. The long normal is not well adapted for defining bed boundaries but efficiently measures $R_t$ in thick beds with moderate invasion. The lateral is not appropriate for defining beds but generally detects thin resistive ones. It also determines $R_t$ in deep invasion if the beds are very thick and fairly homogeneous.

3. The short normal is the best conventional device for estimating $R_t$ in thin beds that are not too deeply invaded and that appear resistive on the lateral. This is true only in formations with moderate resistivities, such as high-porosity sands.

4. In many consolidated sandstones or limestones of medium or low porosity, the invasion is deep enough that the readings of the short devices (short normal and

limestone sonde) are not affected much by the uncontaminated part of the formations. The readings of these devices, after correction for borehole effect by means of departure curves, may provide approximate values for the average resistivity of the invaded zone, $R_i$. $R_i$ is not well defined because it is related to a region where the fluid distribution, and therefore the resistivity, is variable and not well known. Nevertheless, some reasonable assumptions can be made to determine the value of $R_i$. $R_i$ can be used for quantitative analysis when $R_{xo}$ is not provided by the microdevices (low-porosity formations drilled with fresh mud) or when the neutron log does not give the necessary information.

## CONCLUSIONS

The conventional devices are excellent tools for bed definition and correlation in sand-shale series. They provide reliable data for quantitative analyses in this type of formation except in cases of deep invasion, in thin beds, or in beds including thin hard streaks.

In hard formations they are not appropriate for evaluating individual porous and permeable sections. They are, nevertheless, useful for correlation and for determination of average reservoir characteristics, such as porosity and saturation over thick intervals that are predominantly porous and permeable.

For a more complete and accurate definition of the beds and a closer investigation of reservoir characteristics, microdevices and resistivity methods using focusing systems are required. These are described elsewhere in this book.

## SUMMARY - THE SHORT NORMAL CURVE

**Measures:**
- Apparent resistivity; has shallow investigation.

**Features and Applications:**
- A symmetrical curve; peaks in the center of the bed.
- Good results in thin beds.
- A good correlation curve.
- Measures the invaded zone resistivity, $R_i$.

**Limitations:**
- Reverses in beds thinner than 16 in.
- Greatly affected by mud invasion.
- Poor results in salty muds and/or in large-diameter boreholes.

## SUMMARY - THE LONG NORMAL CURVE

**Measures:**
- Apparent resistivity; reads deeper than the short normal.

**Features and Applications:**
- A symmetrical curve; peaks in the center of the bed.
- Good $R_t$ results in beds thicker than 10 ft if invasion is not too deep.

**Limitations:**
- Reverses in beds thinner than 64 in.
- Adversely affected by adjacent formations.
- Not suitable for bed definition.

## SUMMARY - THE 18-ft 8-in. LATERAL CURVE

**Measures:**
- Apparent resistivity; reads deeper than the long normal.

**Features and Applications:**
- Has an asymmetrical curve shape; does not peak in the center of the bed.
- Good $R_t$ results in beds thicker than 30 ft.
- Can detect thin resistive beds.

**Limitations:**
- Not well adapted for bed definition.
- Adversely affected by adjacent formations.
- Reads too low over the top 18 ft 8 in. of a thick bed (called the cutting effect).
- Reads too low for a distance of 18 ft 8 in. below a resistive streak (called the blind zone).

# 5 The Laterolog

## GENERAL INFORMATION

The Laterolog device uses an electrode arrangement and automatic control system to force the current to flow radially through the formations as a sheet of predetermined thickness.

Two types of Laterolog devices were used to record these old logs; the Laterolog-7 and Laterolog-3. The Laterolog-7 tool used "point" electrodes, schematically represented in Fig. 20. The device had one central electrode ($A_0$1) and three pairs of electrodes ($M_1$ and $M_2$, $M_1'$ and $M_2'$, and $A_1$ and $A_2$) placed symmetrically with respect to $A_0$; the pairs being respectively short circuited. A current of constant and calibrated intensity was sent through $A_0$. Additional currents of the same polarity were fed through the auxiliary power electrodes ($A_1$ and $A_2$). The intensity of these currents was automatically and continuously adjusted in such a way that the difference in potential between the short circuited pairs ($M_1M_2$ and $M_1'M_2'$) was maintained practically equal to zero. The potential of any one of these four electrodes was then measured. The apparent resistivity measured with this device was proportional to the value of this potential and to a calibration coefficient that depended on the distance between the electrodes.

With this system, the current emitted from $A_0$ was prevented from flowing upward past $M_1$ and $M_1'$ or downward past $M_2$ and $M_2'$, just as though insulated plugs were placed in the hole at the levels of these electrodes. Accordingly, the distance the current travels across the mud is very small and the mud column has very little influence on the measurements except in the case of deep cavings. Also, the electrical conditions created by the controlling system are such that the current emitted by $A_0$ behaves as if the insulated plugs were extended horizontally some distance away from the borehole. The current is therefore forced to flow within an approximately horizontal slice of space whose thickness is about equal to the distance separating the midpoint ($O_1$) of $M_1M_1'$ from the midpoint ($O_2$) of $M_2M_2'$, which is 32 in.

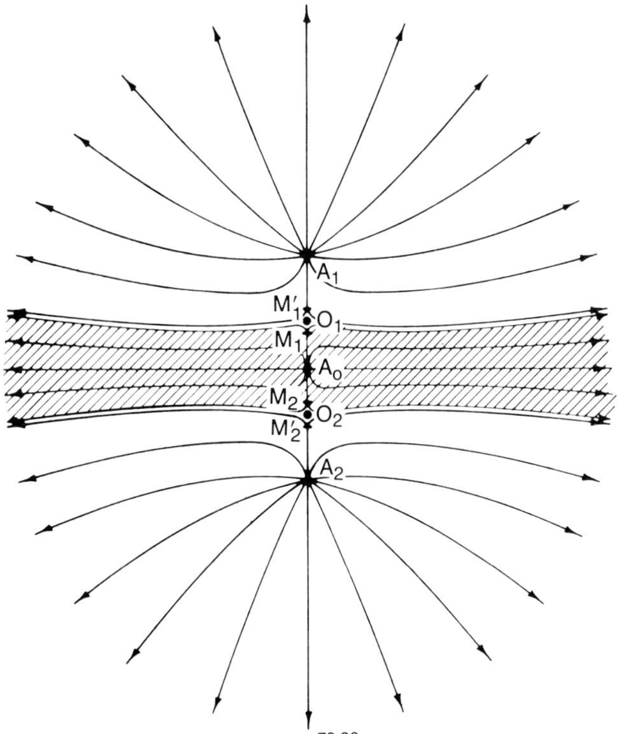


Fig. 20 Distribution of current lines for Laterolog-7.

The Laterolog-3 tool used elongated electrodes, as shown schematically in Fig. 21, and an automatic control system similar to the Laterolog-7 device. Here, also, the current fed through the center electrode ($A_0'$) was forced to concentrate within an almost horizontal slice of space whose thickness was about equal to the distance between $O'_1$ and $O'_2$, which are at the middle of the respective gaps separating the center electrode from the auxiliary electrodes.

If the distances $O_1O_2$ in the Laterolog-7 tool and $O'_1O'_2$ in the Laterolog-3 tool were equal, the thickness of the sheets of currents for both cases would be the same, and the two devices would have similar abilities for detecting details.

The guarded electrode system was similar to the Laterolog-3.

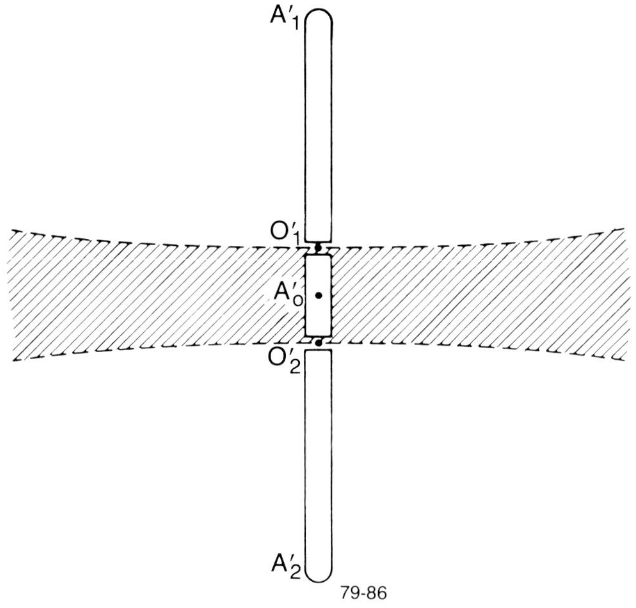


Fig. 21 Laterolog-3 current distribution.

## EFFECT OF BED THICKNESS

The Laterolog device, because of its focusing system, was better adapted for thin beds than the electrical log devices. To illustrate this point, the Laterolog current is compared to the normal device current in a schematic example of a thin noninvaded bed with high resistivity bounded by thick formations of low resistivity. The bed thickness is slightly greater than that of the sheet of current.

Fig. 22 shows this comparison and is divided into two parts by a vertical dashed-dotted line that coincides with the axis of the borehole. A power electrode (A) is shown on the axis midway between the bed boundaries. Distribution of the current emitted from A, without any focusing system (normal device), is represented qualitatively on the left of the figure. The current distribution corresponding to the use of a focusing system (Laterolog) is shown on the right.

The apparent resistivity is proportional to the potential prevailing at an electrode (M) close to the power electrode (A) and, consequently, to the resistance between the equipotential surface passing through M and infinity. Without a focusing system, the current lines diverge from A in all directions and are attracted upward and downward by the adjacent conductive formations so that the resistance offered by the bed is, to a great extent, bypassed. The apparent resistivity read opposite the bed is therefore much lower than the true resistivity of the bed. With the Laterolog device, on the contrary, all current lines flow within the boundaries of the bed, at least over a large distance from the borehole, so that the resistance between the equipotential passing through M and infinity (and, accordingly, the potential at M) is directly controlled by the resistivity of the bed. The apparent resistivity is close to the true resistivity of the bed.

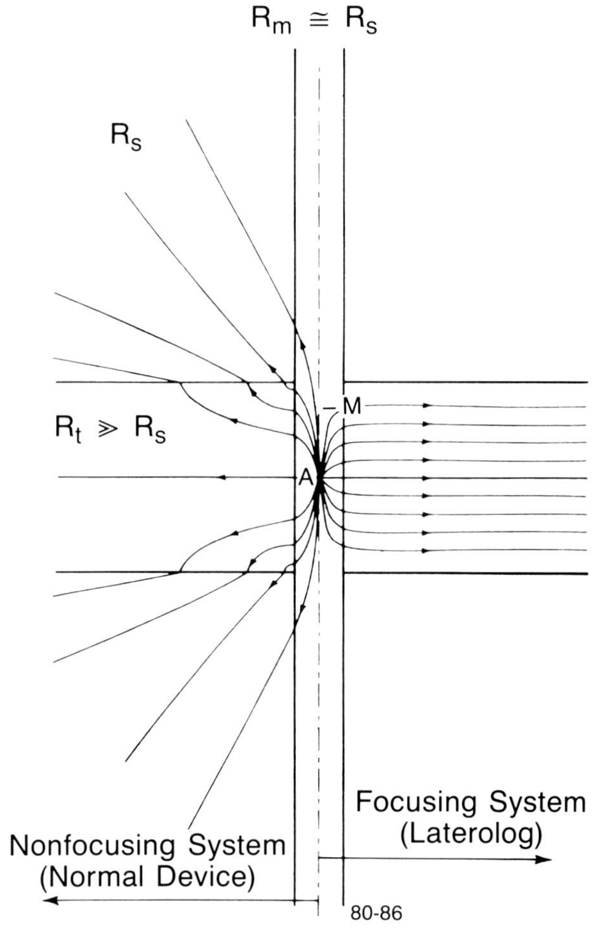


Fig. 22 Comparison of focused vs nonfocused current distribution.

Fig. 23 shows a typical example of a curve recorded in the laboratory opposite a thin noninvaded bed, more resistive than the adjacent formations, and penetrated by a borehole filled with mud of low resistivity.

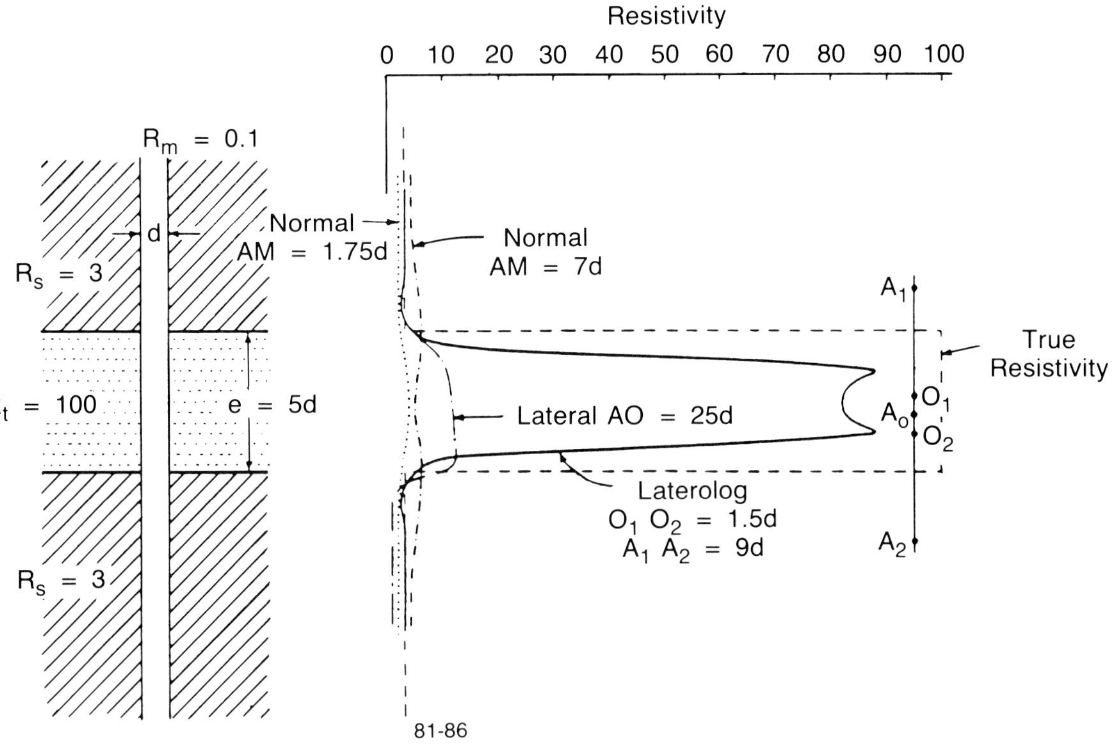


Fig. 23 Comparison of E-log and Laterolog response.

In this figure the bed thickness *e* is equal to five times the hole diameter *d*. The spacing of the Laterolog device is $O_1O_2 = 1.5d$, with $A_1A_2$ equal to six times $O_1O_2$. For a usual hole diameter of about 9 in., these values correspond respectively to a bed thickness slightly less than 4 ft and to a current sheet thickness slightly greater than 1 ft. The true resistivity of the bed is 100 ohm-m, the resistivity of the mud is 0.1 ohm-m, and the resistivity of the adjacent formations is 3 ohm-m.

The electrical log curves recorded with a short normal, a long normal, and a lateral device are shown for comparison.

The Laterolog device gives a much sharper indication of bed boundaries than conventional devices. The "apparent thickness" of the bed, however, is smaller than the true thickness by an amount about equal to the spacing. The value of the apparent resistivity read with the Laterolog on the center plane of the bed is 80 ohm-m, compared to an $R_t$ of 100 ohm-m; the conventional devices show 4, 5, and 12 ohm-m, respectively.

In hard rock the porous and permeable sections are generally sandwiched between tight formations; in other words, $R_t$ is often smaller than $R_s$. Laboratory tests for the case of $R_t < R_s$ show that the effect of adjacent formations can be neglected without too great an error in saturation evaluation if bed thickness is greater than about 4 ft. This conclusion applies to both Laterolog-7 and Laterolog-3 measurements.

**EFFECT OF THE INVADED ZONE**

In permeable beds, the current used for measurement must cross the invaded zone radially before reaching the uncontaminated zone. The apparent resistivity is proportional to a voltage that includes the ohmic drop of potential first across the invaded zone and then across the uncontaminated zone. If the invasion is deep, as it generally is in hard formations, the effect of the invaded zone on the apparent resistivity may be important.

Departure curves have been computed for thick beds, giving the value of the apparent resistivity read with the Laterolog-7 curve versus the true resistivity, for various depths of invasion and various resistivities of the invaded zone.

The computed results are summarized into a few simple equations that correct the readings for invasion effect:

For $d_i = 20$ in.,

$$R_a = 0.2\ R_{xo} + 0.8\ R_t \text{ or } R_t = 1.25\ R_a - 0.25\ R_{xo}.$$

For $d_i = 40$ in.,

$$R_a = 0.4\ R_{xo} + 0.6\ R_t \text{ or } R_t = 1.66\ R_a - 0.66\ R_{xo}.$$

For $d_i = 80$ in.,

$$R_a = 0.6\ R_{xo} + 0.4\ R_t \text{ or } R_t - 2.5\ R_a - 1.5\ R_{xo}.$$

Field experience has shown that the same equations can be applied to the Laterolog-3 readings.

Values for $R_{xo}$ can be determined from the MicroLog or the Microlaterolog; the value of $d_i$ has to be assumed.

Since $R_{xo}$ is a direct function of $R_m$, the effect of the invaded zone depends directly on the mud resistivity. It is interesting to define what the optimum value of $R_m$ should be so that the invaded zone effect is as small as possible.

To illustrate this point the following example is used:

$F = 30$, $R_w = 0.05$ ohm-m, and $d_i = 40$ in.

$R_0$ for 100% water saturation = 1.5 ohm-m.

$R_t$ for 25% water saturation = 24 ohm-m.

Fresh Mud — With a mud resistivity of 1 ohm-m, $R_{xo}$ will be approximately 30 ohm-m. With such a mud, the invaded zone will have little effect when the formation is oil bearing because $R_{xo}$ (30 ohm-m) is close to $R_t$ (24 ohm-m).

For water-bearing formations, with $d_i = 20$, 40, and 80 in., the readings will be equal to 7.2, 12.9, and 18.6 ohm-m, respectively. The apparent resistivities are therefore far greater than the true resistivity ($R_0 = 1.5$ ohm-m) for even a moderate diameter of invasion, such as $d_i = 2d$. In fresh muds the Laterolog is not able to distinguish oil-bearing from water-bearing beds, even qualitatively, except when invasion is very shallow.

Salty Mud — With a mud resistivity of 0.1 ohm-m, $R_{xo}$ is approximately 3 ohm-m.

In the case of oil, the values for $R_a$ are respectively 19.6, 15.4, and 11.4 ohm-m; the true value is 24 ohm-m. When the formation is water bearing, the values of $R_a$ are 1.8, 2.1, and 2.4 ohm-m; the true value is 1.5 ohm-m. The Laterolog curve thus reads below the true resistivity for oil-bearing beds and above for water-bearing beds. The relative errors, nevertheless, are very small for shallow invasion, and, though appreciable, are still acceptable in cases of deep invasion. At any rate, the qualitative discrimination between oil-bearing and water-bearing beds is quite clear.

Conditions are favorable for using the Laterolog tool when the resistivity of the mud is low. The Laterolog is an excellent tool for those regions where the conventional electrical log is the most adversely affected.

Salt muds do, however, have some drawbacks; one of the biggest is that the SP curve is reduced to an almost flat line of no value for discrimination between beds. Taking into account all aspects of the interpretation, the optimum mud resistivity should be from three to five times that of the formation water resistivity. This rule, incidentally, is valid not only for Laterolog conditions but also for electrical logs.

## APPLICATIONS

The Laterolog device gives a sharp, detailed record of the sequences of beds regardless of the mud resistivity and is an excellent tool for formation definition and for correlation. Only the application to reservoir evaluation in hard-rock territories will be discussed because the Laterolog tool is seldom used in other regions.

### High-Salinity Mud

High-salinity muds are frequently found in hard-rock territories because of salt beds overlying the productive formations. The Laterolog curve is one of the components of the so-called "salt mud survey technique," which became standard practice in many of these regions, such as Kansas and the Williston basin. The other components are usually the gamma ray log, the Microlaterolog, and, occasionally, the neutron log.

With this technique, the gamma ray curve makes the distinction between shale beds and nonshale beds; the Microlaterolog delineates, within the nonshale beds, the intervals with enough porosity to be considered potentially productive. The qualitative analysis of each interval so singled out by the Microlaterolog is simple. Since the resistivity of the mud is of the same order of magnitude as that of the formation water, the resistivity of the flushed zone is usually about the same as the true resistivity of the formations when they are water bearing and is smaller than their true resistivity when they contain oil or gas in appreciable quantities. The Laterolog reading is roughly equal to $R_{xo}$ in the former case, and it is greater than $R_{xo}$ but smaller than $R_t$ in the latter case. When the value of $R_{xo}$ from the Microlaterolog is compared with the Laterolog reading, the presence or absence of hydrocarbons can be determined.

On the other hand, although $R_m$ and $R_w$ are of the same order of magnitude, $R_m$ is often great enough with respect to $R_w$ that the SP curve shows deflections of about 10 mV at the junctions between thick sections containing mostly shales or mostly limestones. This SP curve can sometimes be used to determine the average value of $R_w$ over thick intervals when measured $R_w$ values are not available.

### Field Examples

Fig. 24 shows a Laterolog curve recorded in limestones drilled with a salt-saturated mud in west Texas. The figure also includes, for comparison, the resistivity curves recorded with short normal and limestone devices and a gamma ray log. One striking feature is the considerable difference in resistivities recorded with the various devices.

The resistivities recorded with the short normal and the limestone sonde are very low, not greater than about 10 ohm-m, because of the small radii of investigation of these devices and the bypass effect of the mud column. In comparison, resistivities as high as 150 ohm-m are obtained with the lateral device (radius of investigation is very large) and the Laterolog.

Many details on the Laterolog curve are lacking on the limestone sonde curve, despite the resistivity scale for the limestone curve being five times as sensitive as that used for the Laterolog recording.

Although large resistivity contrasts are observed on both the lateral and Laterolog curves, the shapes of the curves are markedly different. The comparison shows clearly how much the lateral curve is distorted by the effect of the mud column. In particular, the curve fails to show thin highly resistive streaks, such as A, that correspond to sharp peaks on the Laterolog curve.

Fig. 25 is an example of the salt mud survey technique in a sequence of limestones and shales in Kansas. Interval A is a potential reservoir. Over this interval, the average value of $R_{xo}$ is equal to the average reading of the Microlaterolog; i.e., 10 ohm-m.

The Laterolog reading is 38 ohm-m. Furthermore, $R_{mf}$ at BHT is 0.07 ohm-m and $R_w$ is known from local experience to be 0.04 ohm-m. With 20% residual oil saturation in the flushed zone, a lower limit for water saturation is 31% and the average porosity is 10%.

This interval was perforated, and after treatment with 8,000 gal of acid produced 10 barrels of oil per hour with 5% water.

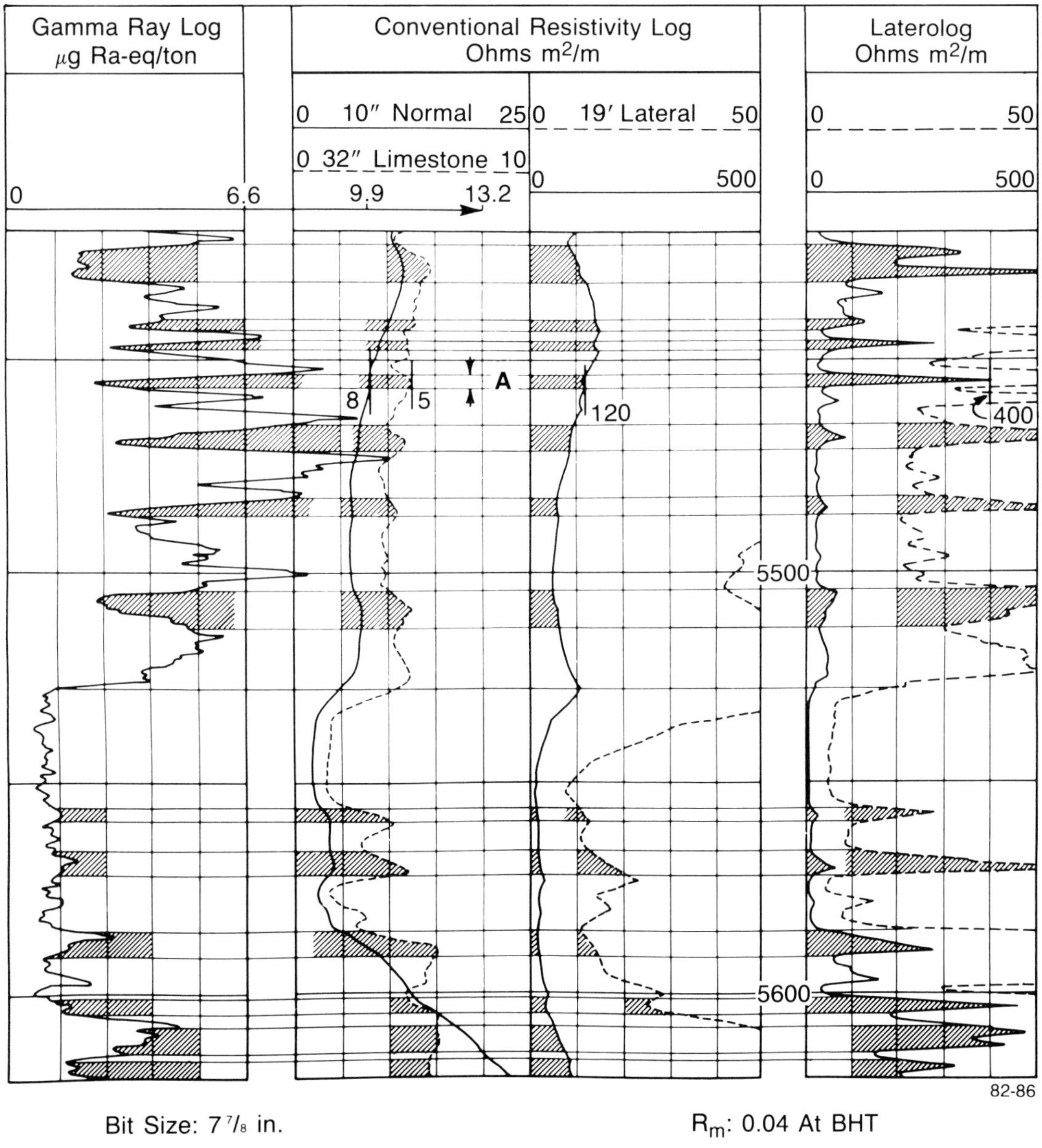


Fig. 24 Comparison of conventional resistivity log, Laterolog, and gamma ray log in hard formations with salt mud (Permian basin).

The Microlaterolog device measures $R_{xo}$ directly in this situation. (The Microlaterolog measurement is covered in Chapter 7).

## SUMMARY - THE LATEROLOG

**Measures:**

- Apparent formation resistivity.

**Features and Applications:**

- A focused measurement with sharp bed definition.
- Accurate at high resistivities.
- Best results in salt mud systems.

**Limitations:**

- Inaccurate at low resistivities in fresh muds.

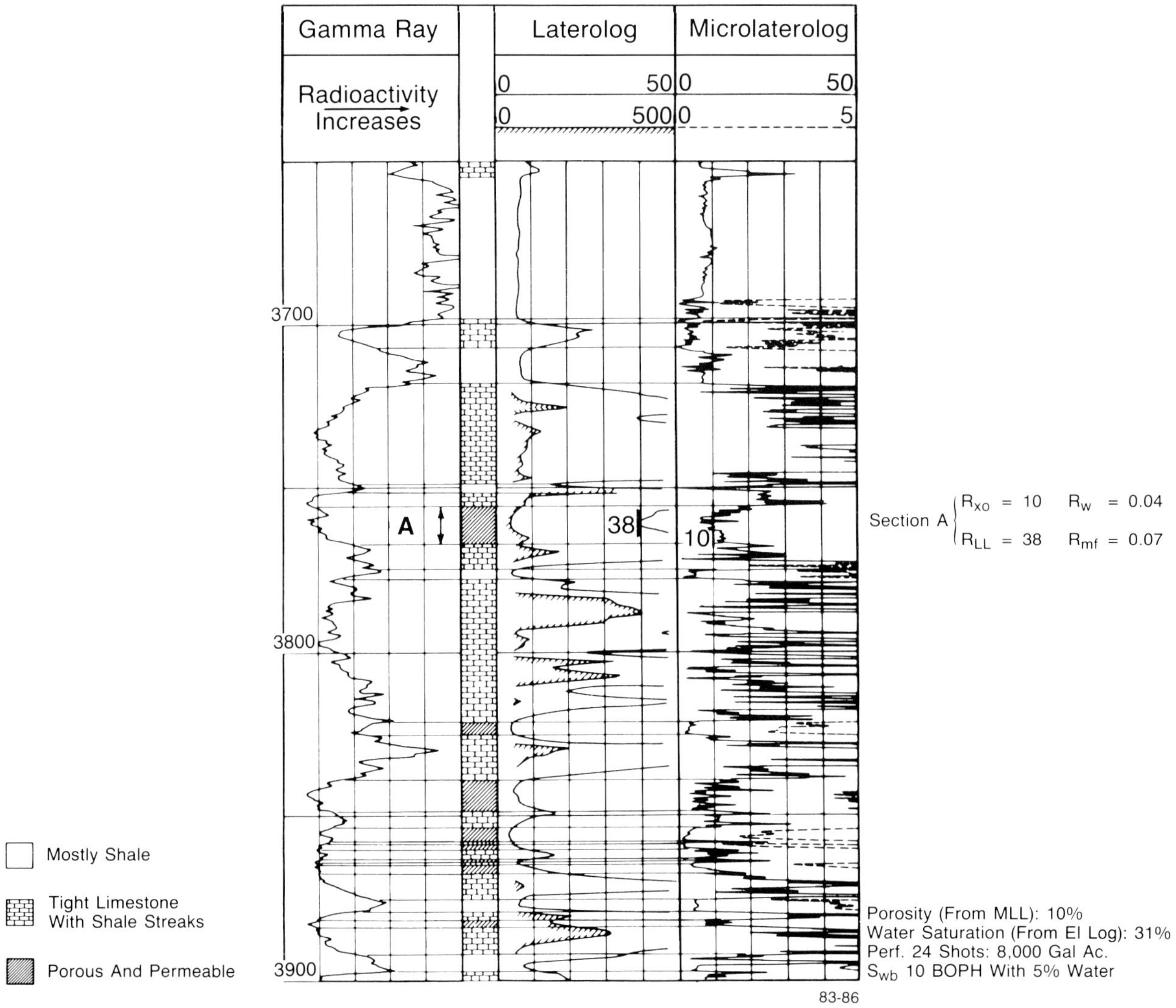


Fig. 25 Typical salt mud survey.

# 6 The MicroLog

## GENERAL INFORMATION

Because the SP curve does not accurately delineate permeable beds in hard or appreciably consolidated formations, the MicroLog (ML) tool was developed. It can be used in moderately consolidated and soft formations drilled with fresh muds for precise delineation of beds, evaluation of sand count, and determination of porosity.

The MicroLog sonde (Fig. 26) contains three electrodes spaced at 1-in. intervals along a vertical line in the middle of a rubber pad. The pad is attached to a movable spring-actuated arm. When the tool is in the hole, the spring presses the pad against the wall of the hole with a pressure that is practically constant over a range of hole diameters. The pressure of the pad causes the electrodes to directly contact the formation. The pad shields them from the short-circuiting effect of the mud, and the system measures the resistivity of the small volume of formation located in front of the pad.

The three electrodes on the pad, in conjunction with the surface electrode, make it possible to record two curves simultaneously. A constant current is sent from the lower electrode (A). The potential occurring between the two other electrodes ($M_1$ and $M_2$) is used to record a microinverse curve. At the same time, the potential appearing between $M_2$ and the surface electrode is used to record a micronormal curve. Since the volume of the formation measured by the inverse electrodes is smaller than the volumn measured by the normal electrodes, the inverse curve is more affected by conditions near the borehole wall; that is, the microinverse has a smaller radius of investigation than the micronormal.

A second spring-actuated arm diametrically opposed to the first keeps the tool centered in the hole. Since the separation of the two arms is proportional to the diameter of the hole, a caliper device was added. The microcaliper curve is recorded simultaneously with the other two curves and shows variations of hole diameter as small as ⅛ in.

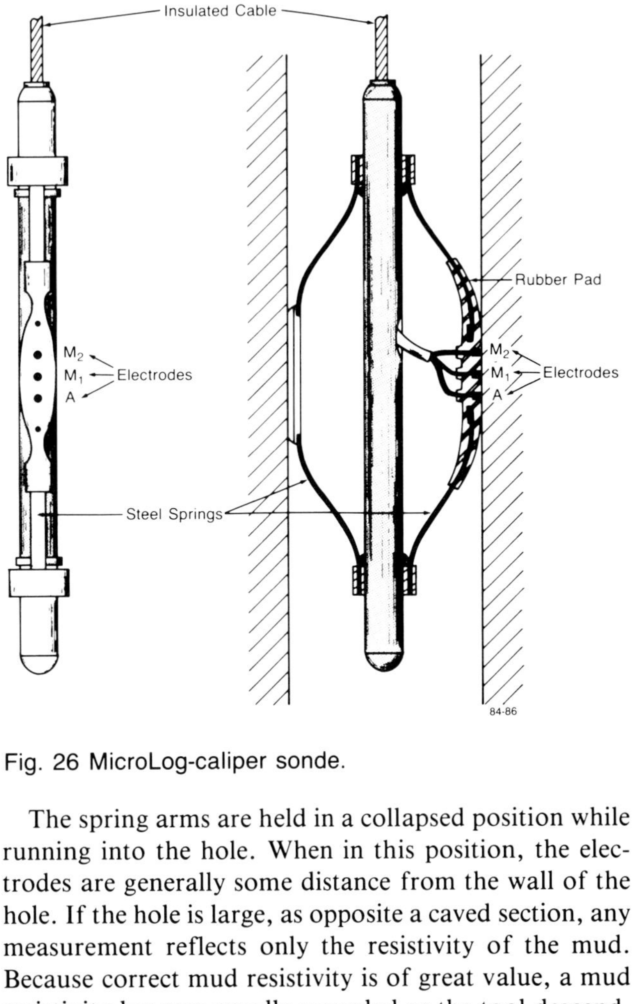


Fig. 26 MicroLog-caliper sonde.

The spring arms are held in a collapsed position while running into the hole. When in this position, the electrodes are generally some distance from the wall of the hole. If the hole is large, as opposite a caved section, any measurement reflects only the resistivity of the mud. Because correct mud resistivity is of great value, a mud resistivity log was usually recorded as the tool descended. The MicroLog curves are recorded with the arms open as the tool comes out of the hole.

The old MicroLog-caliper survey consisted of four curves: microinverse, micronormal, microcaliper, and mud resistivity.

## DELINEATION OF DIFFERENT FORMATIONS

### Porous and Permeable Beds

If a borehole filled with mud is drilled through a porous and permeable bed, invasion will occur and mudcake will be deposited on the walls of the hole opposite the bed. Mud filtrate will almost completely replace the original formation fluids for several inches into the bed immediately adjacent to the borehole. This flushed zone resistivity is directly related to that of the mud. The resistivity of the 3 or 4 in. of material adjacent to the wall of the hole, constituted of mudcake and invaded formation, will have a relationship to the mud resistivity but will seldom exceed it by more than a factor of 20. Because of its small depth of investigation, the microinverse device is influenced by the mudcake more than is the micronormal. Therefore, opposite a permeable and porous bed the two curves usually show different values of resistivity. The difference between the values is referred to as separation.

In most cases the resistivity of the flushed zone is greater than that of the mudcake. When this is true, the micronormal gives a higher apparent resistivity than the microinverse, and the log is said to have "positive separation."

When the resistivity of the flushed zone is greater than the resistivity of the mudcake, a bed can be interpreted as porous and permeable if 1) there are comparatively low resistivity readings, not more than 20 times the mud resistivity; and 2) if there is positive separation between the microinverse and micronormal curves.

An exception to the second rule occurs when the mudcakes are very thin. This is usually in high-salinity muds, and the separation is practically negligible.

When the invasion is very shallow and the formation is saltwater bearing, the depth of invasion may be so small that the volume of material involved in the resistivity measurements includes a portion of the formation beyond the flushed zone. Under such conditions the mudcake can be more resistive than the formations immediately behind, and the micronormal reading will be lower than the microinverse; in other words, the separation is negative. Shallow invasion can be observed in high-porosity formations when low-waterloss mud is used. When the porosity and vertical permeability are very high, as in many places in the Gulf Coast, shallow invasion usually occurs.

The mudcake effect smoothes out the resistivity readings so that the curves seldom show sharp variations opposite a permeable bed, even when the formation factor changes appreciably within the bed.

The same criteria apply for granular porosity and for secondary porosity consisting of a thin network of vugs and fissures.

The MicroLog device can detect permeable beds even if the permeability is very low (down to a fraction of 1 md), but there must be sufficient permeability to allow the buildup of mudcake. It may happen, therefore, that a bed indicated as permeable by the MicroLog may have a permeability too low for commercial production or for any production at all.

### Tight Sections

In tight sections the electrode system is separated from the formation only by a very thin mud film. The MicroLog readings are very high, at least 20 times the mud resistivity. Furthermore, because the wall of the hole is not perfectly smooth, the thickness of the mud film is not constant and therefore its influence is not constant. As a result, the curves show numerous sharp peaks and depressions; the separation may be positive or negative.

### Shales

When there is no caving in the shales, the pad is separated from the formations by only a thin mud film. The reading is equal to or, more often, smaller than the resistivity of the shales. The separation may be negative, zero, or slightly positive. This behavior of the MicroLog curve opposite shales has not been entirely explained. It has been assumed that a negative separation may be partly due to the anisotropy of the shales.

In practice, when a positive separation is observed opposite a shale, it may be confused with a permeable bed and other curves can be used to resolve the ambiguity. In many cases, the SP curve and/or the gamma ray curve suffice. Often, however, these curves do not give clear enough indications opposite very thin beds, particularly in hard formations. The microcaliper is extremely helpful since a decrease in hole diameter is likely to correspond to a mudcake on the permeable bed. An increase in hole size usually indicates shale.

### Cavings

Cavings are most often encountered in shale beds, but they may also occur opposite other types of formations, including hard rocks. In the case of cavings, the pad generally does not make good contact with the wall. The fact that the MicroLog curves read a resistivity equal to the mud resistivity is usually a conspicuous indication of caving. In many cases, however, the interpretation is not so easy. If the cave is not too deep, the readings may still be affected by the formations. Since the mud is usually less resistive than the formations, a positive separation may be interpreted as a permeable streak.

### Hard Formations

Fig. 27 shows a section of mostly limestones and dolomites of Permian age in west Texas. The different permeable zones are clearly indicated on the MicroLog by low resistivities with positive separation. Numerous peaks obviously correspond to compact sections. A few shale beds are also apparent. These are characterized by low resistivities, absence of separation, and deflections of the SP curve toward the positive side.

Some large negative anomalies of the SP curve, which might be interpreted as corresponding to permeable formations, actually occur opposite beds that are shown to be tight by the MicroLog curves (see Sections a and b). The porosity of these beds, although indicated as very low by the MicroLog readings, is large enough that SP currents can circulate.

Conversely, deflections of the SP curve toward the positive side are also noticeable opposite beds that, according to the MicroLog values, are compact. In fact, very thin streaks within hard formations may cause these positive deflections on the SP curve. This is clearly apparent in sections such as c and d, where the MicroLog shows a series of very sharp ripples that indicate this interbedding of shales and limestones. It may also be that argillaceous material mixed with the limestones causes the SP curve to deflect toward the positive side.

Fig. 28 includes a microcaliper log. The SP curve indicates Interval A is permeable. The MicroLog recording shows that it is constituted of thin interbedded permeable and tight sections. The exact boundary of each individual streak is clearly shown, and net pay can be determined.

Each section indicated as permeable by the MicroLog curves is confirmed by the microcaliper, which shows a decrease in the hole diameter whenever mudcake is present.

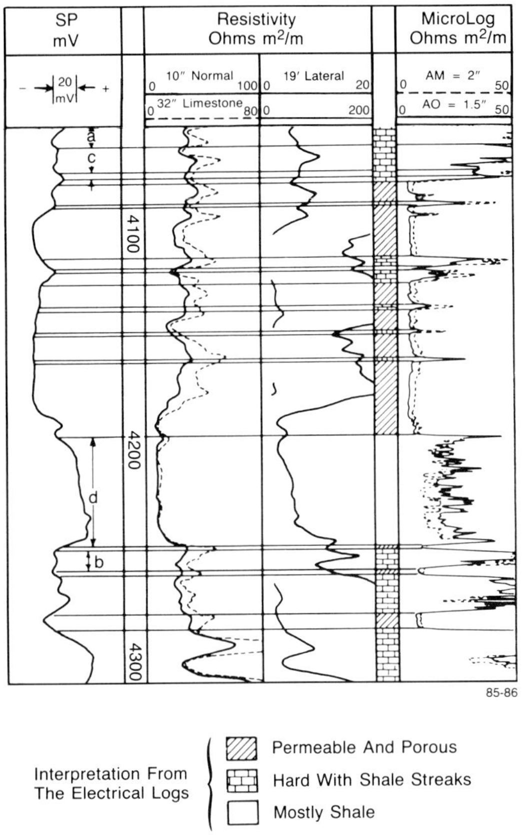


Fig. 27 Example of MicroLog in limestone formations (west Texas).

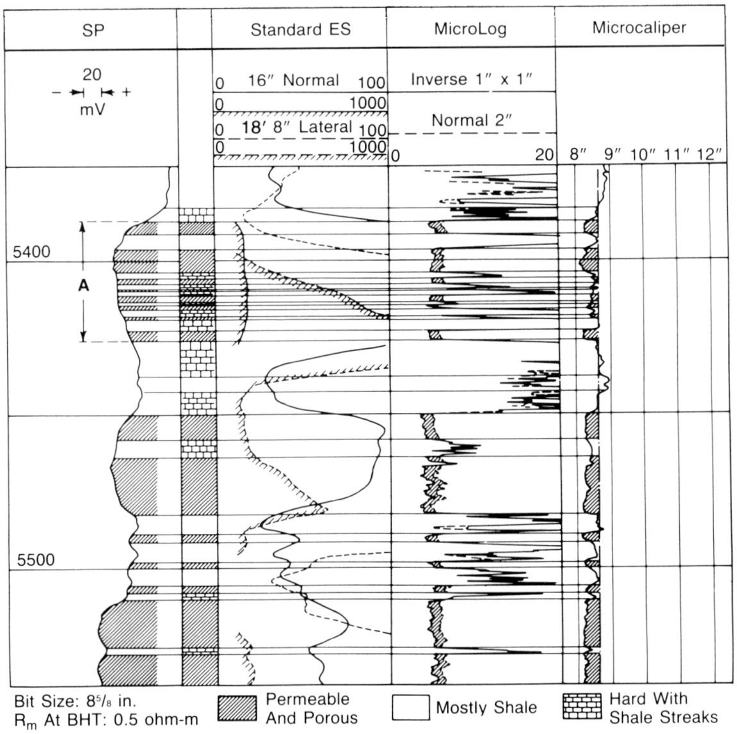


Fig. 28 Example of MicroLog in hard formation (Permian basin).

**Sand-Shale Series**

Fig. 29 is an example of conventional electrical log and MicroLog curves in a Gulf Coast well. In the interpretation column, boundaries are indicated between beds belonging to three formation types:

1. Shales are indicated by low MicroLog readings with a minimum of separation between the curves. Shale and sand laminations are further characterized by breaks in the microcurves.

2. Unconsolidated sands are shown by somewhat higher readings than shales, usually with a positive separation between curves.

3. Consolidated streaks correspond to high MicroLog values. Such streaks have a very low permeability and should be eliminated in the determination of net pay sections.

There is a difference in the resistivities recorded with the 2-in. micronormal opposite oil-bearing and saltwater beds. In this instance, the water level is at approximately 9410 ft. Above this depth, the average reading of the 2-in. micronormal is 2.7 ohm-m. Below this depth, the average reading is only 2 ohm-m. This feature occurs frequently. When the lithologic characteristics are constant and the connate water salinities are approximately constant throughout a series of sands, a difference is observed on the MicroLog readings depending on whether the sands contain oil. Oil is not completely flushed by mud filtrate, and this causes a relatively higher resistivity in the flushed zone. It may have been caused by the very small depth of the flushed zone and the MicroLog readings may have been affected by the uninvaded part of the formation.

Cases of very shallow invasion are often observed in unconsolidated sands. Fig. 30 illustrates the behavior of the logs, in particular the MicroLog, in a sequence of shales and sands of high porosity and permeability. The sands are all saltwater bearing and the drilling mud is appreciably fresher than the connate water.

Section A can be divided into Parts a and b according to the electrical logs and MicroLog curves. In Part a, both short and long normal curves track together, which indicates a very shallow invasion. The MicroLog shows a negative separation. In Part b, just below the tight adjacent bed, the greater values recorded by both indicate a deeper invasion. This is confirmed by the higher values and the positive separation observed on the MicroLog curves.

Similar considerations apply to other permeable sections. The feature is slightly different in Bed B. The lower part, with shallow invasion, is clearly differentiated from the upper part with deeper invasion. However, in the lower part, the short normal curve stays higher than the long normal, which shows invasion is not as shallow as in Zone a in Bed A. Correspondingly, the separation between MicroLog curves, instead of being negative, is slightly positive.

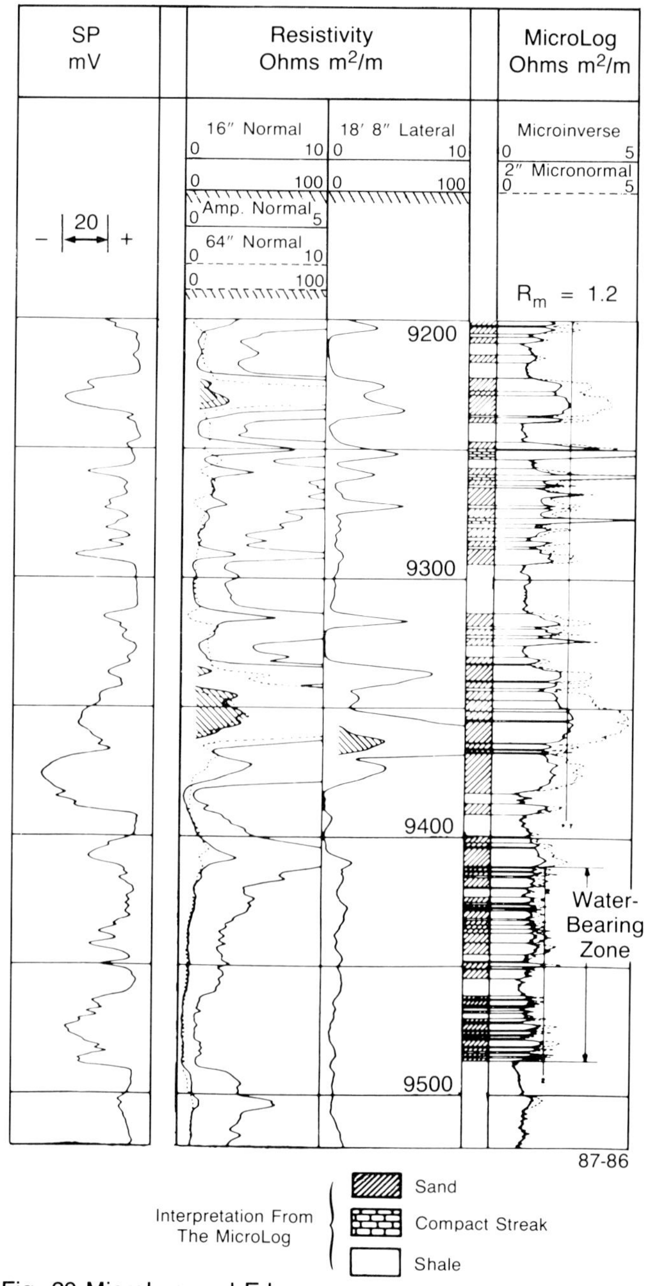


Fig. 29 MicroLog and E-log.

## QUANTITATIVE ANALYSIS

In water-bearing formations a flushed zone is saturated with mud filtrate only. If the formation is clean, the formation factor can be considered equal to $R_{xo}/R_{mf}$, with $R_{xo}$ being the resistivity of the flushed zone and $R_{mf}$ the resistivity of the mud filtrate.

As already stated, in oil-bearing formations residual oil saturation in the flushed zone affects the measurements, and a correction is necessary. The proportion of residual oil can be determined from core analyses, but it is generally accepted that a value of 20% does not usually entail large errors, at least in formations with granular porosity. Greater residual oil saturations are common, however, in the cases of high-gravity oil. Gas-bearing formations may also display high residual saturations in the flushed zone.

$R_{mf}$ can be measured directly on samples of filtrate pressed out of the mud by a filter press. If this cannot be done, $R_{mf}$ can be estimated from the value of $R_m$, the mud resistivity, by a chart based on average statistical laboratory data. The evaluation of the formation factor and, therefore, the porosity depends on the determination of $R_{xo}$.

Except in cases of very shallow invasion, the radius of investigation does not extend beyond the flushed zone. The measured resistivities are functions of the flushed zone resistivity ($R_{xo}$), the hole diameter ($d$), the mudcake resistivity ($R_{mc}$), and the mudcake thickness, $t_{mc}$. Interpretation charts have been established, based on laboratory tests, which make it possible to derive $R_{xo}$ and $t_{mc}$ from the two MicroLog readings when $d$ and $R_{mc}$ are known. The hole diameter is generally known from the microcaliper log or is estimated from bit size. The value of $R_{mc}$ can be obtained by direct measurements on mudcakes pressed from actual samples of the mud or from a chart based on average statistical data. MicroLog interpretation charts are available for logs recorded with the old hard-rubber WRS-D pad and for logs recorded later with the hydraulic pad.

When $R_{xo}$ is not too great compared to $R_{mc}$, MicroLog readings are not much affected by the mudcake and $R_{xo}$ can be obtained with fair accuracy. This is the case in high porosity. The contrary situation occurs when $R_{xo}$ is very great in comparison to $R_{mc}$; i.e., in the case of low porosity. This effect of the value of $R_{xo}/R_{mc}$ is shown clearly by the interpretation charts. For high values of $R_{xo}/R_{mc}$, curves for constant values of $R_{xo}/R_{mc}$ are very crowded so that a small error in the determinations of $R_{2''}$ and $R_{1''x1''}$ corresponds to a much larger error in the value of $R_{xo}$. There is, therefore, a lower limit of porosity values below which the quantitative interpretation of the MicroLog readings is not reliable. With the hydraulic pad, this lower porosity limit is about 10%.

## FIELD EXAMPLES

The first example, Fig. 31, shows the electrical log and MicroLog curves recorded through the Yegua formation in a Gulf Coast well. The bit size for this well is 9⅞ in. and $R_m$ is 1.2 ohm-m at BHT (150° F).

The averaged values of porosity computed from the logs for the three sand intervals (1, 2, and 3) are shown on the figure. These are matched with the average values derived from the core analysis. In this study Intervals a, b, c, and d have been excluded because all logs indicate they contain substantial amounts of shale.

The second example, Fig. 32, shows the MicroLog and electrical log curves recorded in a north Texas well.

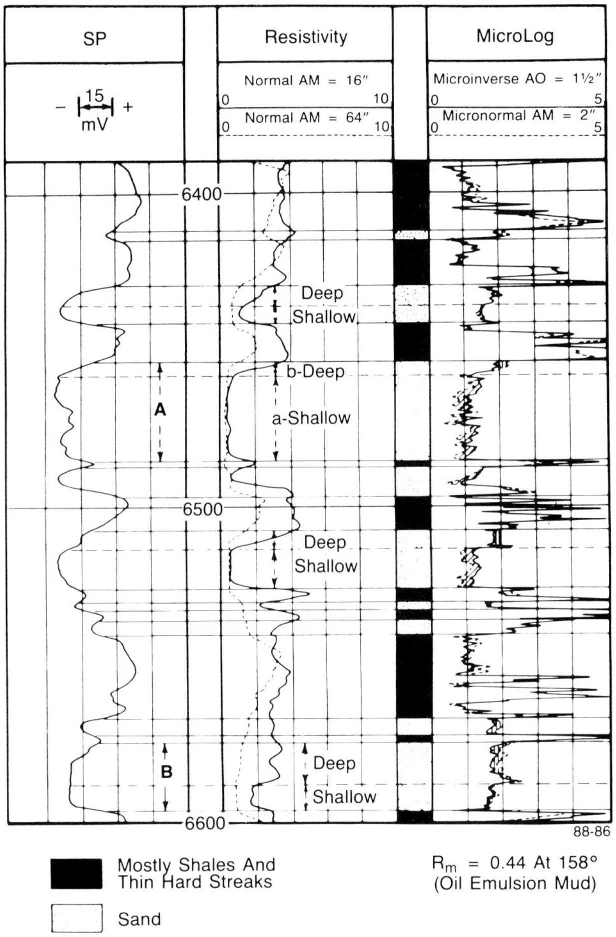


Fig. 30 MicroLog and E-log: shallow invasion in saltwater sands.

Several porous and permeable sections are indicated by the MicroLog curves (crosshatched areas on the figure).

If the thin streak denoted as "a" is discarded, the quantitative analysis of the MicroLog curves in Section A shows that the formation factor and the porosity vary within comparatively narrow limits throughout the section. In other words, the whole section can be regarded as one fairly homogeneous bed. The total thickness is about 70 ft; so the effect of the adjacent formations on the conventional curves, even on the long lateral curve, can be neglected.

The average value of $R_{xo}/R_{mf}$ throughout the section, as derived from the MicroLog charts, is 40. The average formation factor, using 15% residual oil in the flushed zone, is about 30. The corresponding value of the average porosity is about 17%.

The amplitude of the SP curve is equal to −90 mV. With $K$ equal to 80 at BHT, the corresponding $R_{mf}/R_w$ is about 14. A value of $R_m$ = 0.07 ohm-m at BHT corresponds to a value of 0.5 ohm-m for $R_{mf}$. $R_w$ is thus about 0.035 ohm-m.

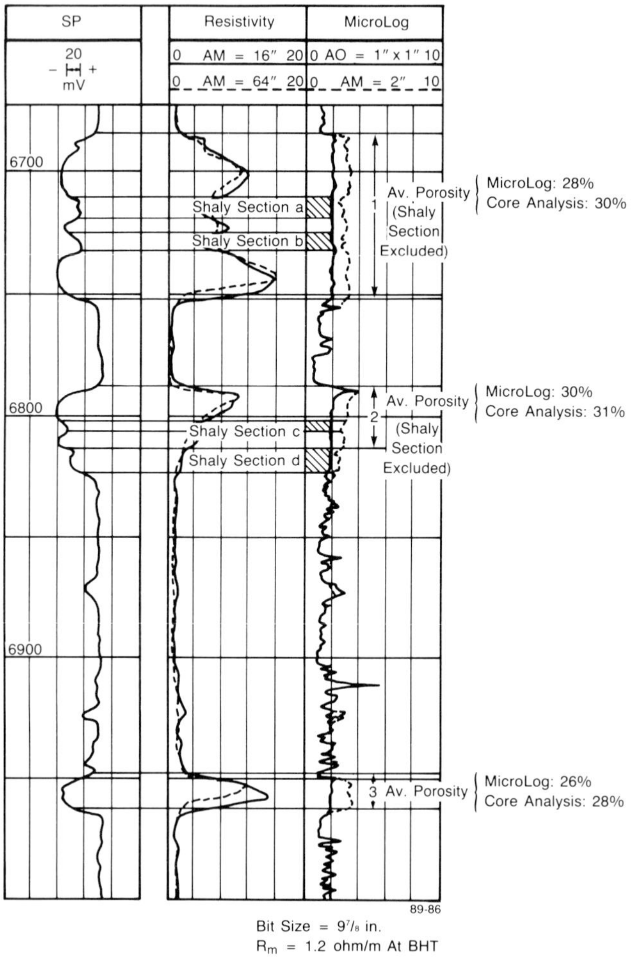


Fig. 31 Example of quantitative interpretation of MicroLog in clean formations of high porosity (Gulf Coast).

The average apparent resistivity read on the long lateral throughout the section is equal to 220 ohm-m, which, after correction with the departure curves, gives an average value of 130 ohm-m for $R_t$.

The water saturation in the uncontaminated zone is therefore

$$S_w = \left(\frac{30 \times 0.035}{130}\right)^{1/2}, \text{ or } 9\%.$$

The computations of water saturation in the other permeable, porous sections are not so easy because these sections are comparatively thin and their true resistivities cannot be determined accurately from the electrical logs. For example, considering Section B, whose thickness is about 4 ft according to the MicroLog reading, it appears that the long lateral curve is very distorted and unreliable. Because of the tight formations above and below the bed, it is likely that the value read on the long normal (100 ohm-m) is an upper limit for $R_t$. A reasonable lower limit is 50 ohm-m. Furthermore, the SP curve is not entirely developed; it can be assumed that the value of $R_w$ is the same as in Section A, 0.035 ohm-m. The MicroLog interpretation gives a formation factor value of 30; therefore, the water saturation would fall between $(35 \times 0.035/100)^{1/2}$ or 11%, and $(35 \times 0.035/50)^{1/2}$ or 15%. This result is, of course, a very rough approximation because of the uncertain values involved, particularly $R_t$.

Almost all the permeable sections were perforated; the well produced 214 barrels of oil per day through a 16/64-in. choke.

## CONCLUSIONS

The MicroLog device can discriminate between very thin beds having different lithological characteristics because of its very small electrode spacings and because the insulated pad shields the electrode system from the mud column. Its resolving power is of the order of a few inches so it shows extremely sharp excursions at the boundaries between beds.

With the help of the SP or gamma ray curve and the microcaliper reading, the MicroLog curves can be used to differentiate between porous, permeable formations and impervious beds (hard rocks and shales). It gives the exact location of the boundaries of potential reservoirs and makes it possible to compute the "net pay."

With appropriate interpretation charts, the MicroLog curves can give the value of the resistivity of the flushed zone in formations of high or medium porosity. From this value, the formation factor of substantially clean formations can be estimated. The MicroLog device is quite valuable for determining porosity in clean formations.

With all other conditions being the same, $R_{xo}$ differs with varying hydrocarbon or water content. Accordingly, in formations of comparatively high and constant porosity where the micronormal reading is not too different from $R_{xo}$, it can show the location of the oil-water or gas-water contacts.

Several MicroLog interpretation charts are available for determining formation factor and porosity. An intermediate step in the process involves the determination of $R_{xo}$. These charts are found in Section C of Appendix A. There were basically two different styles of MicroLog measuring pads: the hydraulic pad and the older solid rubber, or D-type, pad. A separate chart is needed for each type pad.

An empirical MicroLog interpretation method is also found in Section C. This method will give porosity directly without first finding $R_{xo}$. Its application is widely used and surprisingly accurate.

## SUMMARY - MICROLOG

**Measures:**

- Resistivity of a small volume of formation just behind the borehole wall.

**Features and Applications:**

- Defines very thin permeable beds in hard rock.
- Provides accurate sand count in soft rock.
- Provides estimates of formation factor and porosity.

**Limitations:**

- Poor results in large or caved holes.
- Inaccurate at low porosities.
- Fresh mud system required.

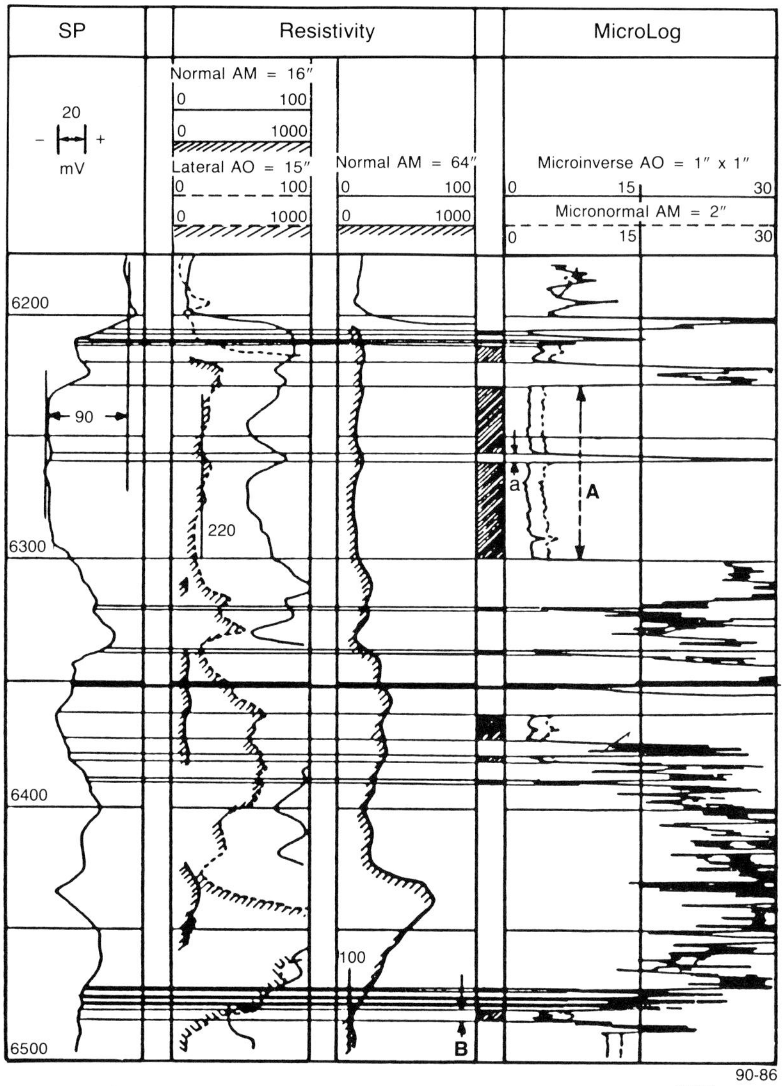


Use Of MicroLog
To Determine Porosity And Fluid Saturation

Section A:

| | |
|---|---|
| Average Value Of $\frac{R_{xo}}{R_{mf}}$ From MicroLog | = 40 |
| Average Porosity (Supposing 15% ROS) | 17% |
| Average Formation Factor | 30 |
| $R_w$ From SP Curve | 0.035 |
| $R_t$ From Lateral | 130 |
| Water Saturation: $\sqrt{\frac{30 \times 0.035}{130}}$ | 10% |

Bit size: $7^7/_8$ in.

$R_m$ (At BHT) = 0.7 ohm/m

Porous And Permeable

Fig. 32 Interpretation of E-log and MicroLog.

# 7 The Microlaterolog

## GENERAL INFORMATION

When the porosity of a formation is less than 10%, the survey current on the MicroLog device flows mostly in the mudcake and the measurements reflect little of the character of the formation. For this reason, the MicroLog is not the proper tool for an accurate determination of $R_{xo}$ and, therefore, of formation factor and porosity under these conditions. Another tool was developed to give satisfactory results when the formations are of relatively low porosity. The Microlaterolog (MLL) tool, introduced in 1952, is a microdevice that uses a focusing system to reduce the effect of the mudcake on the measurement. As long as the mudcake is thin its effect is negligible, and the Microlaterolog device can be used to determine formation factor and porosity in all types of formations and in wells with high-salinity muds since, as a rule, these have thin mudcakes.

The Microlaterolog device has one very small center electrode ($A_o$) and three circular electrodes ($M_1$, $M_2$, and $A_1$); each of the three ring electrodes is concentric with $A_o$. The spacing between the electrodes varies from ½ to 1 in. (Fig. 33). These electrodes are imbedded in an insulating pad that is applied against the borehole wall by a spring system.

A current of constant and known intensity is sent through the center electrode ($A_o$); another current of the same phase is emitted from the outer ring electrode ($A_1$). The intensity of the latter current is automatically and continuously adjusted in such a way that the difference of potential between $M_1$ and $M_2$ is kept practically at zero. The potential then existing at either $M_1$ or $M_2$ is measured. The apparent resistivity recorded is proportional to a value of this potential and to a factor that depends on the geometrical characteristics of the system. This factor is determined experimentally. Since the action of the currents from $A_o$ and $A_1$, respectively, is to keep $M_1$ and $M_2$ at the same potential, there must be no net current flow between $M_1$ and $M_2$. If this is the case, the current from $A_o$ must be constrained to a beam that flows horizontally outward toward the formation. At the wall of the hole, the thickness of the beam is approximately equal to the average diameter of $M_1$ and $M_2$; i.e., 1 to 2 in. The thickness of the beam increases slowly at first and then rapidly until, at a distance of approximately 3 in. beyond the wall of the hole, it has become so broad that the formation beyond that point has little effect on the potential of M and N and, therefore, on the resistivity measurement.

In a porous and permeable formation, the electrode system is separated from the formation by the mudcake. Since the current crosses the mudcake in a direction perpendicular to the wall, the distance it has to travel

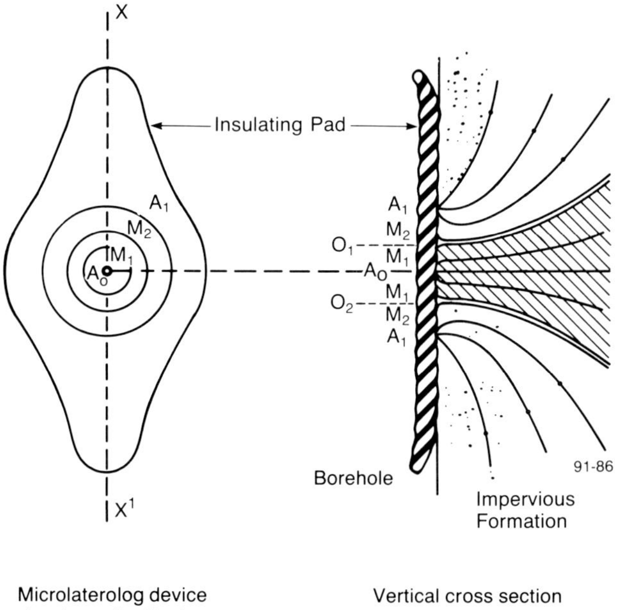


Microlaterolog device showing distribution of electrodes

Vertical cross section showing the current lines

Fig. 33 Microlaterolog focusing system.

across the mudcake is small in comparison with the length of the path through the formation; furthermore, the resistivity of the mudcake is usually less than the resistivity of the formation. As a result of these two factors, the influence of the mudcake on the measurement is much reduced and may be negligible if the mudcake thickness is small. Fig. 34 illustrates that the mudcake constitutes a path toward the mud column for a part of the current from the MicroLog tool; it generally affects the measurements much more than with the Microlaterolog device, particularly in formations with low porosity.

Both laboratory tests and field experience have proved that the Microlaterolog equipment provides a measurement of $R_{xo}$ and, therefore, a value of formation factor and porosity without correction when the mudcake thickness is less than about 3/8 in.

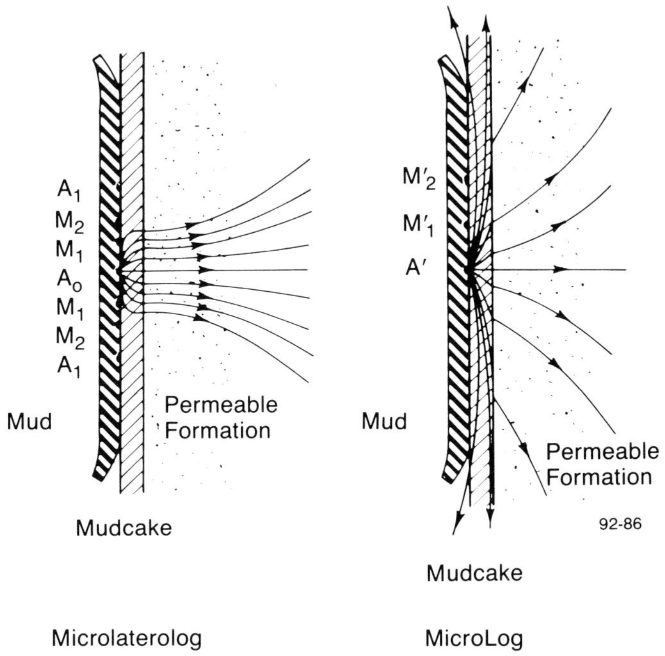


Fig. 34 Comparative distribution of current lines for Microlaterolog and MicroLog.

## FIELD TECHNIQUE

The Microlaterolog equipment is shown in Fig. 35. Electrodes $A_o$, $M_1$, $M_2$, and $A_1$ are molded into the rubber pad. The shape of the rubber pad and the pressure exerted by the spring are such that the pad fits the wall of the hole over a substantial area around the electrodes. The electrodes are connected by insulated wires to the electronic cartridge.

Fig. 36 shows a schematic representation of the circuits used for recording a Microlaterolog curve. The source of alternating current is used to send a current of predetermined and constant intensity into $A_o$. Electrodes $M_1$ and $M_2$ are connected to the input terminals of an automatic control system. Electrode $A_1$ is connected to one of the output terminals of this system; the other output terminal is grounded through a body of the cartridge. With this system, a current is sent through $A_1$, which continuously acts to maintain a zero difference of potential between $M_1$ and $M_2$. The common potential of $M_1$ and $M_2$ is rectified and recorded at the surface. A microcaliper is recorded simultaneously with the Microlaterolog curve.

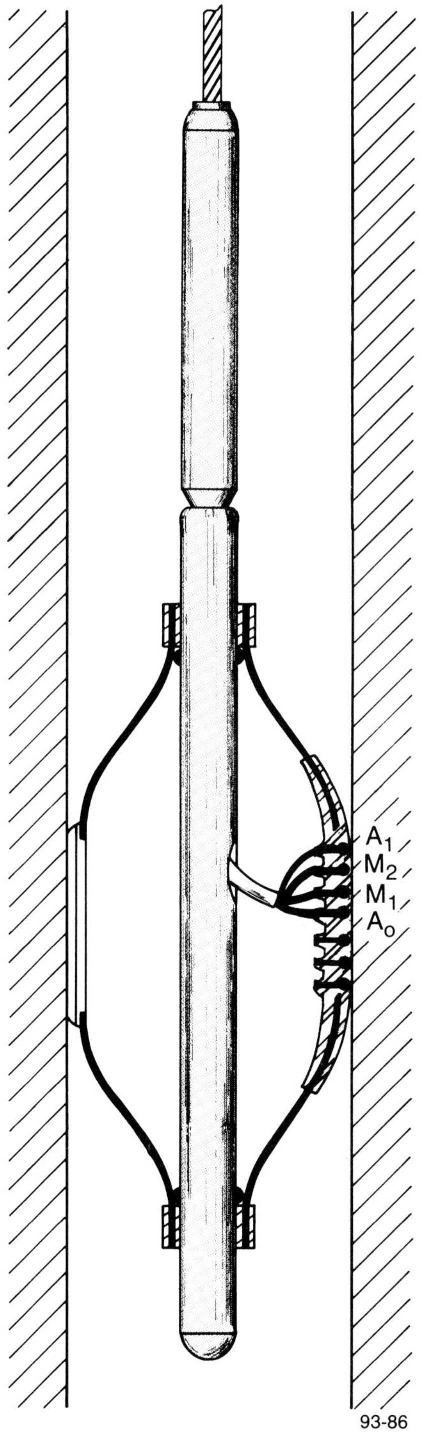


Fig. 35 Schematic representation of Microlaterolog equipment.

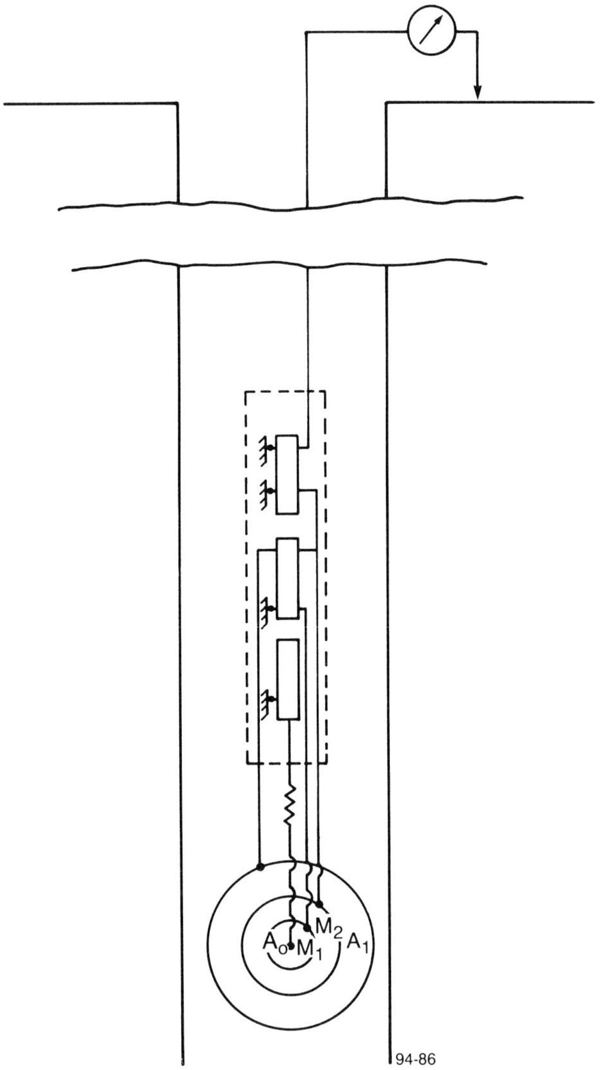


Fig. 36 Schematic representation of Microlaterolog circuits.

## INTERPRETATION AND APPLICATIONS

Since the current beam has a very small diameter (about 2 in.), the depth resolution of the Microlaterolog device is very high and the log gives a very detailed definition of beds, down to approximately 2 in. in thickness. Because of this extreme detail, the logs were often recorded with two depth scales; the regular one (5 in. for 100 ft) for correlation and an enlarged one (25 in. for 100 ft) for detailed study.

### Tight Formations

Since these formations are usually in gauge, the pad is well seated and high resistivities are measured.

### Shales

When there is no caving, the pad makes good contact with the formation and the tool measures the resistivity of the shale. As in the case of the MicroLog device, cavings may give features to the curve similar to those for porous beds. Again, the discrimination of these features is possible with the help of the SP, gamma ray, and microcaliper curves.

### Porous and Permeable Formations

The Microlaterolog tool was designed to measure $R_{xo}$ — a valuable parameter for quantitative analysis with electrical logs. When mudcake thickness is less than ⅜ in. the log measures $R_{xo}$ directly without correction.

The Microlaterolog, one of the components of the salt mud survey technique, was used extensively in hard-rock territories where high-salinity muds are common (Kansas, Michigan, Permian basin, and Williston basin).

## FIELD EXAMPLE

Fig. 37 shows an example of Microlaterolog and Laterolog curves recorded in a well drilled with high-salinity mud in Pennsylvanian limestone in Oklahoma. $R_w$ was 0.04 ohm-m and $R_{mf}$ was 0.05 ohm-m at formation temperature.

The Microlaterolog curve gives an extremely sharp delineation of the beds. In this case the resistivity measured should be very close to $R_{xo}$.

The low Microlaterolog reading indicates a zone of interest from 6106 to 6113 ft. The Microlaterolog curve reads an average value of about 7 ohm-m. The apparent formation factor $F_a = R_{xo}/R_{mf} \times (1 - \text{ROS})^2 = 7/0.05 \times 0.64 = 75$. This represents a porosity of about 11.5%. At this interval, $R_t$ = 28 ohm-m from the Laterolog. To calculate water saturation by the ratio method ($R_{xo}/R_t$ and $R_{mf}/R_w$) use, respectively, 0.25 and 1.25. Entering Chart D-4 with these values gives $S_w$ = 35%. The chart assumes average $S_{xo}$ values. The equation to solve for $S_w$ would be

$$S_w = S_{xo} \left( \frac{F\, R_w}{R_t} \right)^{1/2}.$$

Substitution gives a value of 33% for $S_w$. Remember that $S_{xo} = (1 - \text{ROS})$, where ROS is the residual oil saturation in the flushed zone.

The assumptions made in the example were 1) that the Microlaterolog curve was reading $R_{xo}$ directly; 2) that the Laterolog curve was reading $R_t$; and 3) that the ROS was 20%.

The well produced 418 barrels of oil per day after an acid fracture job.

**SUMMARY - MICROLATEROLOG**

**Measures:**

- Resistivity of a small volume of formation just behind the borehole wall.

**Features and Applications:**

- A focused measurement with very sharp bed definition in all formations.
- Very accurate $R_{xo}$ determination.
- Determines formation factor and porosity.
- Good results in salty mud system.

**Limitations:**

- Poor results in large or caved holes.
- Inaccurate in thick mudcakes (reads too low).

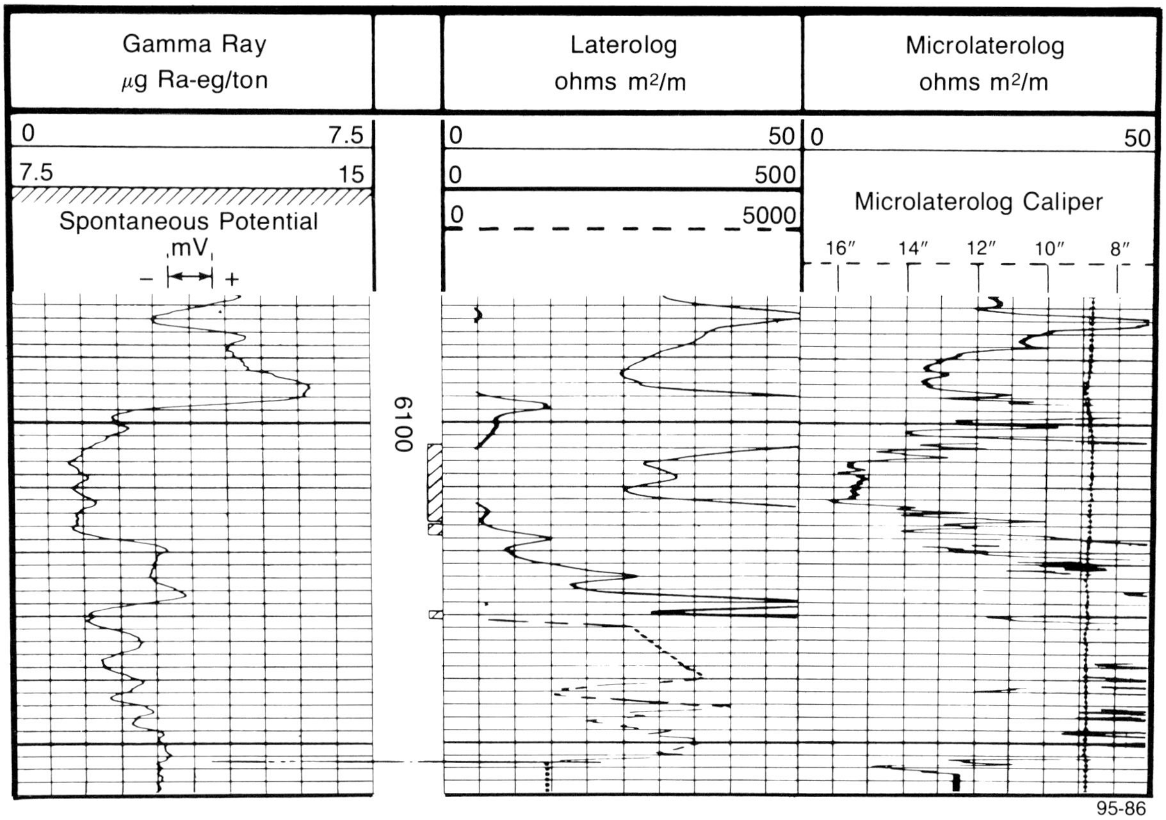


Fig. 37 Field example of Laterolog and Microlaterolog.
Courtesy of Douglas W. Hilchie *Old Electrical Log Interpretation* (1979).

# 8 The Limestone Lateral Curve

No text on old logging applications would be complete without an explanation of the limestone lateral curve. Out of desperation to detect porous layers within massive carbonates, Marcel Schlumberger designed a double lateral sonde (Fig. 38). The assembly, being symmetrical, eliminated the most serious shortcoming of the lateral curve.

The ability of this tool to detect and define conductive beds, while greater than that of other conventional devices, is exceeded considerably by that of the MicroLog or Microlaterolog tools, which came later. Actually, its most important feature was that it often helped to estimate porosity.

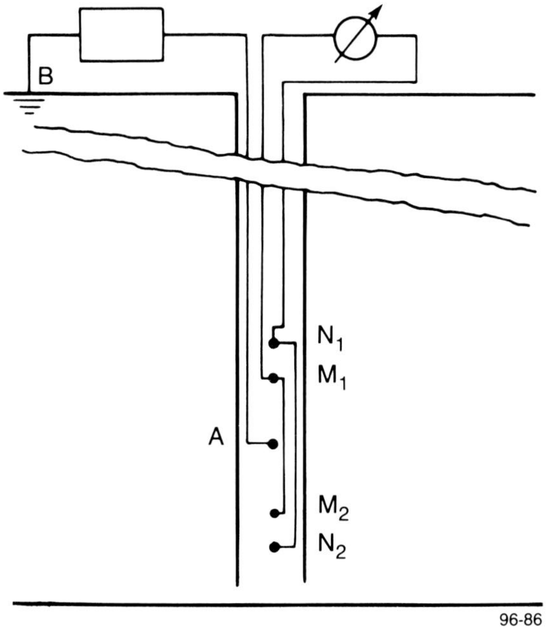


Fig. 38 Limestone sonde.

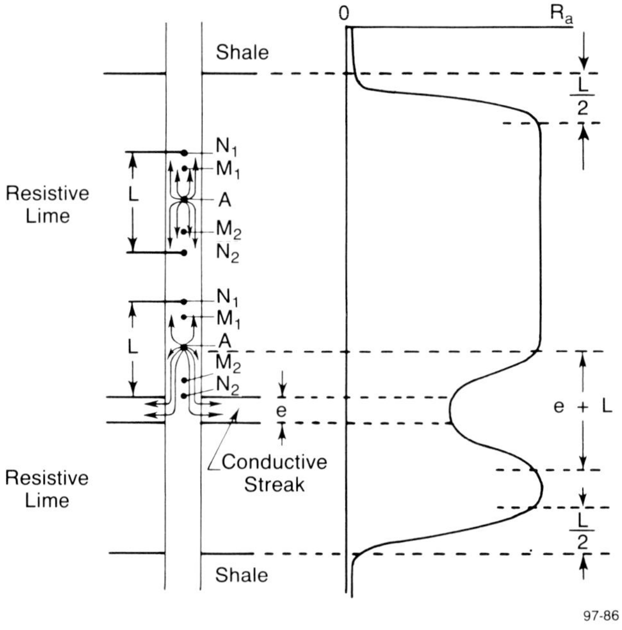


Fig. 39 Schematic drawing illustrating principle of limestone sonde.

In Fig. 39 the apparent resistivity recorded by the limestone sonde is constant so long as all its electrodes are totally within the highly resistive bed (top part of the figure). Furthermore, the maximum value of resistivity measured ($R_a$) depends only upon the hole size and the mud resistivity. The reading starts to decline when the device encounters a conductive bed. A "ball park" value of porosity can be determined with this measurement.

## FIELD EXAMPLE

Most of the limestone logs were run in the Permian basin between 1945 and 1950. Fig. 40 is a typical west Texas electrical survey recorded during this period. The maximum resistivity recorded with the limestone device is 82 to 83 ohm-m. Zones of interest are labeled A, B, and C because they exhibit lower resistivities than the maximum value and because they are located within the massive bed between 9640 and 9725 ft.

An interpretation chart for the limestone lateral is shown in Fig. 41. To solve for a porosity index (PI), enter the chart with a value corresponding to the percent of maximum deflection and the hole size. For Zone A, the percent deflection $^{57}/_{82}$ x 100 = 69%; the hole size is given as 8¾ in. The PI is 8%. We can refine this answer by making the corrections outlined in the tables at the bottom of the chart. Zones B and C have slightly higher porosities because the limestone curve shows lower values.

The limestone curve, when scaled in porosity units, is nearly logarithmic so it correlates well with the count rate neutron log.

In some cases, other resistivity measurements can also be related to porosity changes in highly resistive formations. The 10-in. normal curve responds to porosity changes in a similar manner to the limestone curve. For Zone A, the short normal reads about 66 ohm-m. It must, however, be corrected for borehole effects ($R_a/R_m$ = 66/0.75 = 88). This value is entered on Chart B-4 (AM spacing is 10 in.), and the value of $R_i/R_m$ for an 8¾-in. hole is about 200. Chart C-16 can then be used to convert the value of $R_i/R_m$ into a porosity index. Entering the value of $R_i/R_m$ = 200 gives a porosity of around 7%. Similar procedures with Zones B and C yield porosity values slightly higher, around 8%. This method should be used only as a last resort.

The limestone device was discontinued when the MicroLog device became available.

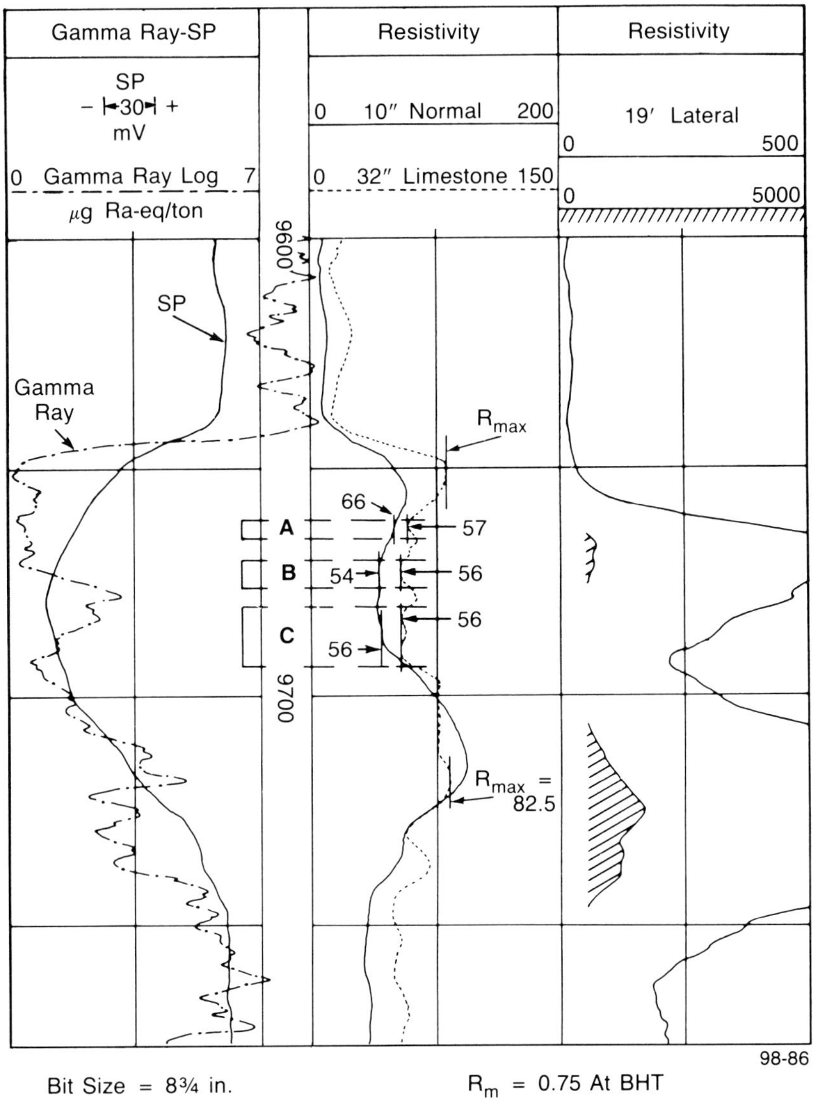


Fig. 40 Field example of E-log with limestone curve.

## SUMMARY - LIMESTONE LATERAL

**Measures:**

- Apparent formation resistivity using a double lateral arrangement that is symmetrical.

**Features and Applications:**

- Can detect porous and permeable layers within massive carbonates.
- Provides a porosity index in low porosities.
- Generally measures in the invaded zones ($R_i$).
- Useful in correlation in massive carbonates.

**Limitations:**

- Inaccurate at low resistivities.
- Adversely affected by salty muds.

POROSITY INDEX FROM LIMESTONE LATERAL

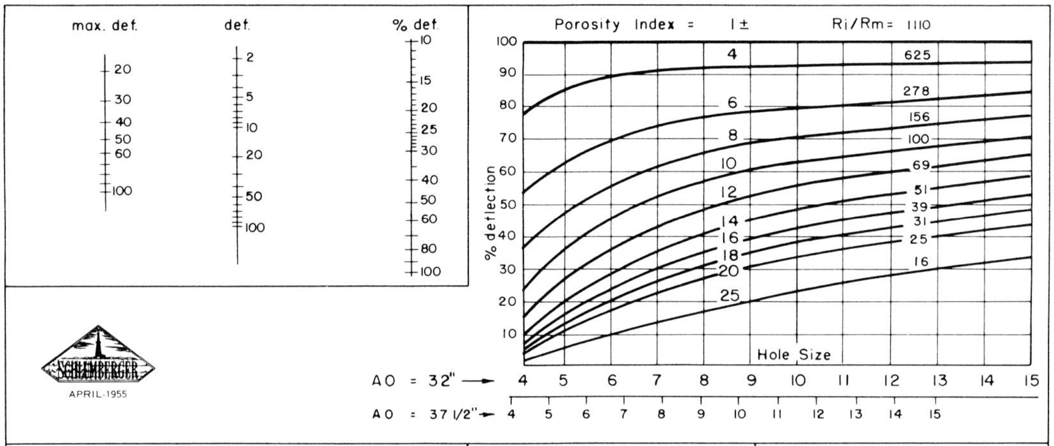


| Max. Ra/Rm | | |
|---|---|---|
| HOLE SIZE | AO = 32" | AO = 37 1/2" |
| 4 3/4" | 445 | 605 |
| 6 3/4" | 177 | 232 |
| 7 7/8" | 128 | 171 |
| 8 3/4" | 102 | 141 |

| WATER - BEARING FORMATIONS | | OIL - BEARING FORMATIONS | |
|---|---|---|---|
| WHEN | THEN | WHEN | THEN |
| " SP = 0 | " C = 1 | " SP = 0 | " C = 1.4 |
| " SP = -50 MV | " C = 0.86 | " SP = -50 MV | " C = 1.22 |
| " SP = -100MV | " C = 0.7 | " SP = -100MV | " C = 1 |

99-86

Fig. 41 Porosity index from limestone lateral.

# 9 Radiation Logging

## THE GAMMA RAY LOG

The gamma ray log is a measurement of the natural radioactivity in the formations. Gamma rays are spontaneously emitted by a few elements; the most common are uranium, potassium, and thorium. It is these gamma rays that are detected in gamma ray logging.

Shales contain the greatest concentration of radioactive salts and are generally much more radioactive than sands, limestones, or anhydrites. There are a few radioactive dolomites that contain a significant amount of uranium. There are also a few radioactive sands. Productive formations such as the Granite Wash found in Texas, New Mexico, and Oklahoma can also contain significant amounts of natural radiation.

Generally, a shale will contain several of the radioactive elements, and it is impossible to tell, with the old gamma ray tool, whether the gamma radiation is primarily due to one or another of these elements. Present-day tools that can break out the percentages of thorium, uranium, and potassium are called natural gamma ray spectrometry tools.

Basically, the GR log distinguishes shales from other formations. The recording equipment was designed so that the curve deflects toward the right when the radioactivity increases. Therefore, the GR curve is somewhat similar to the SP curve, for which deflections to the right also indicate shales.

A gamma ray log can be recorded in open or cased hole, either empty or filled with any type of fluid. In open holes this log is used principally for bed definition and correlation. In holes drilled with very salty mud where the SP lacks resolution, it will locate shales and shaly formations. It is one of the best single-measurement shale indicators. In cased holes, a gamma ray log is used primarily for correlation with openhole logs.

### Unit Of Radioactivity

Today these logs are recorded in API units. Clean sands and carbonates generally display levels of radioactivity from 15 to 20 API units, and shales display from 120 to 240 API units. Most logs recorded prior to 1958 were calibrated in terms of ''radium-equivalent''per ton; this unit was commonly used for many years among geologists. Radium is one of the intermediate elements in the step-by-step disintegration of uranium; consequently, there is a fixed uranium/radium ratio in the earth. Schlumberger used the unit micrograms of radium equivalent per metric ton (1 = 16.5 API units). Other logging companies used different units, such as counts per second.

Many of the early gamma ray logs were made with very inefficient ionization chambers and Geiger-Mueller detectors. Scintillation detectors did not appear on the scene until the late 1950's. Consequently, many of the early logs had high statistical variations and lacked formation detail.

The fundamentals of gamma ray log interpretation are illustrated on Fig. 42. Section B, where the radioactivity is high, is mainly shale. Sections A and C, with lower levels of radioactivity, are mostly nonshales; i.e., sands, sandstones, and limestones.

Large deflections of the curve can be interpreted with little or no ambiguity. Small deflections, however, may be caused by small variations in shale content or by statistical fluctuations. This can be particularly troublesome in regions where radioactivity contrasts between shales and nonshales are small (Section B).

### Field Example

Fig. 43 shows a gamma ray curve recorded in open hole and an SP curve recorded in the same series of sands and shales. Direct correlation is easy. At 9119½ ft, the SP log shows a sharp contrast between shale and sand. The excursion of the gamma ray curve, at the same level, is very rapid at first and rounds as the minimum deflection is approached. A similar feature is observed at 8996 ft. The sharp break at 8960 ft also illustrates the behavior of the gamma ray curve when the tool passes the top of a sand.

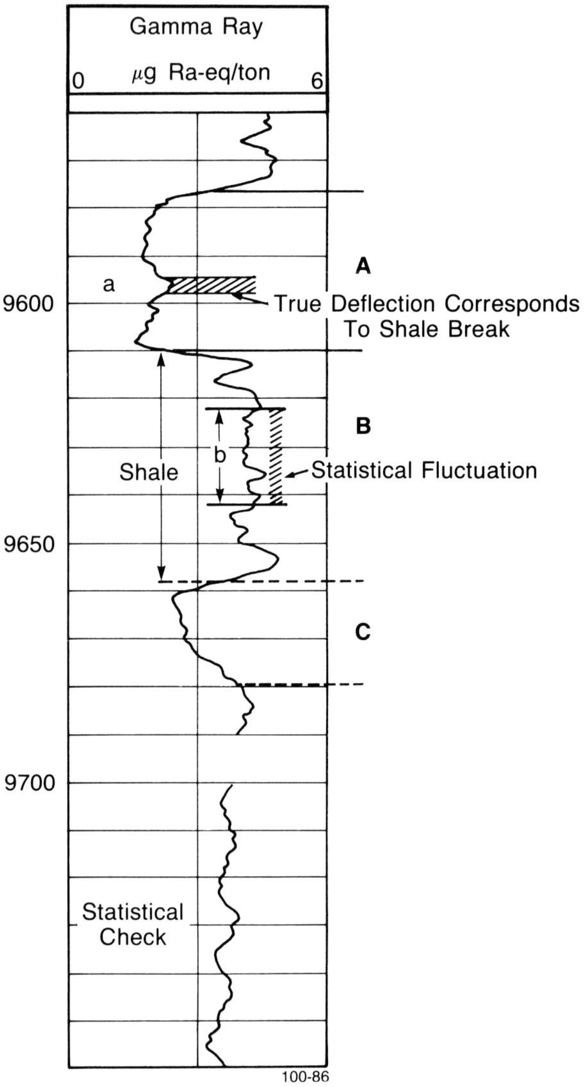


Fig. 42 Typical gamma ray curve.

## SUMMARY - THE GAMMA RAY LOG

**Measures:**

- The formation's natural gamma radiation.

**Features and Applications:**

- A correlation curve similar to the SP.
- Good resolution in any environment — air, salt water, oil — open or cased holes.
- A good shale indicator.

**Limitations:**

- Slow logging speed.
- Statistical variations.
- Sensitive to large or caved holes.

## THE NEUTRON LOG

### General Information

The neutron is one of the elementary particles of the atom. Its mass equals that of the hydrogen atom, and it is electrically neutral. When used in well logging, neutrons are emitted at relatively high velocities from a special source. The neutrons collide with other atomic nuclei as they travel from the source into the formation. Each collision results in a loss of energy or a slowing of the neutron to the thermal velocity, where it is absorbed by the nucleus of a hydrogen atom. This absorption results in the emission of a high energy gamma ray, which is called a "capture"gamma ray.

The logging system consists of a source of fast neutrons and a detector, which are placed close together (about 15 in. apart) in a sonde. If the detector is sensitive to capture gamma rays, the corresponding logs are called neutron-gamma logs. Most early neutron logs were of this type.

The number of gamma rays counted by the logging detector is inversely proportional to the hydrogen concentration in the vicinity of the borehole. Since the presence of hydrogen is almost solely due to the formation fluids the amount of hydrogen per unit volume of rock is a function of its porosity. The porosity can be evaluated by counting the number of gamma rays of capture if the pores are filled with oil or water. The same is not true, however, of gas, since the hydrogen concentration is much lower. A gas-bearing zone will therefore appear on the neutron log as less porous than if it were filled with oil or water.

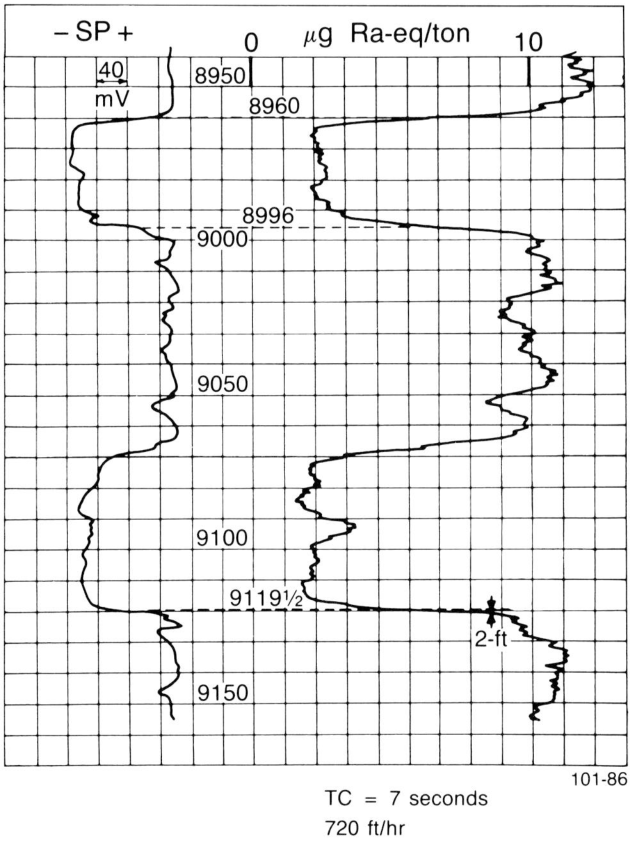


Fig. 43 Comparison of SP and gamma ray curves.

**Qualitative Interpretation**

The old count rate neutron log was calibrated in standard counts per second. If the borehole conditions were reasonable, intervals on the neutron log with a lower counting rate will correspond to higher porosity and/or to greater shale content. With the help of the SP and gamma ray curves, the log deflections that correspond to variations in porosity can be differentiated, to a great extent, from those corresponding to variations in shale content.

Fig. 44 shows a gamma ray and neutron log run through a sequence of sands and shales in an open hole filled with water-based mud. Zones A and B can be interpreted as shaly sands; in the lower section below 8790 ft there are mostly clean sands with porosity variations indicated by the neutron log.

The neutron log in hard formations is shown in Fig. 45, together with a MicroLog and SP curve. Sections A, B, and C are characterized by low neutron readings, which indicate porosity. Section D is a sequence of very thin hard streaks and shale breaks. Section E is a limestone. This interval seems to be a thick bed with low average porosity. The boundaries of the thin beds are not well defined by the neutron curve.

Observations made in the early 1950's in the Permian basin provided practical data on porosity determinations from the neutron log. A linear relation between the logarithm of the porosity and the neutron log reading was observed in many cases. Consequently, the neutron log became an important porosity measurement.

The neutron log gives porosity directly, without the intermediate step of finding the formation factor.

**Effect of Borehole Conditions**

In many cases, empirical approaches can be used to derive fairly reliable porosity values from the neutron logs, at least in open hole and in clean formations. Appreciable errors, however, may occur because of the influence of borehole conditions. For accuracy, the borehole effects should be taken into account. The following conditions have considerable influence on the neutron reading:

- diameter of the open hole
- increasing mud salinity
- presence of casing and cement
- empty holes with a diameter greater than 6 in.
- tool standoff

**Porosity Determination**

The earliest Schlumberger gamma ray-neutron tools were 3⅝-in. sondes, types GNAM 1 to 5. This designation will appear in the log heading. Later tools were called GNT-F or G equipment. These tools used RaBe or PuBe radiation sources. The standard spacing between detector and source was 19½ in., but there were some logs recorded with a 15½-in. spacing. Before porosity determinations can be made, it is very important to know which tool and which spacing were used. The log heading provides this information.

It was common to calibrate the response of the neutron log in porosity values for each given geological unit by empirical comparisons between the deflections of the neutron curve and the porosity values given by core analyses in a sufficient number of wells.

Another very practical interpretation technique to find porosity from the neutron curve is presented in Charts C-17 and C-18 in Appendix A. This technique can also be applied graphically in the following manner: using a logarithmic grid or scaler, mark off a porosity index scale

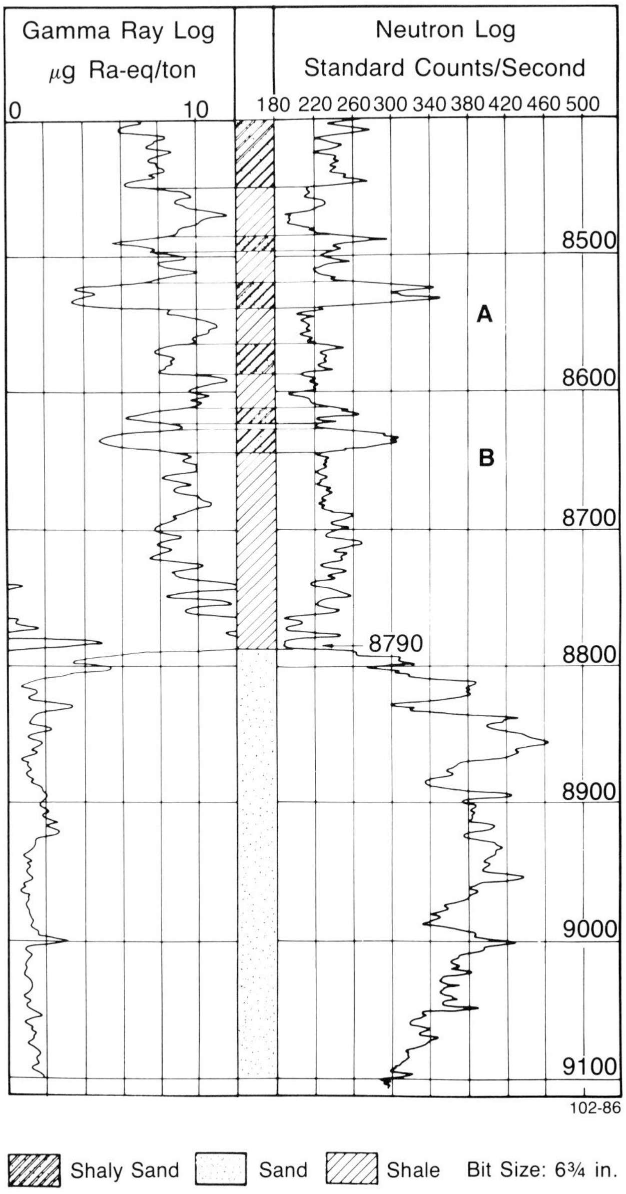


Fig. 44 Gamma ray-neutron log.

from 1 to 50%. Place the scaler on the log with the 1% value at a known tight lime streak. Place the 40% point of the scaler at a shale section. Porosity values for zones of interest can then be estimated using the logarithmic scale. Obviously, if you don't have the proper control points, the technique cannot be used.

## SUMMARY - THE NEUTRON LOG

**Measures:**

- Capture gamma rays - proportional to the hydrogen concentration of the formation.

**Features and Applications:**

- A good correlation curve.
- Determines formation porosity except in gas.
- A good gas detector.
- Works in air, water, and oil environments.
- Works in open or cased holes.

**Limitations:**

- Slow logging speed.
- Sensitive to borehole size, casing and cement thickness, and tool standoff.
- Pronounced lithology effect.

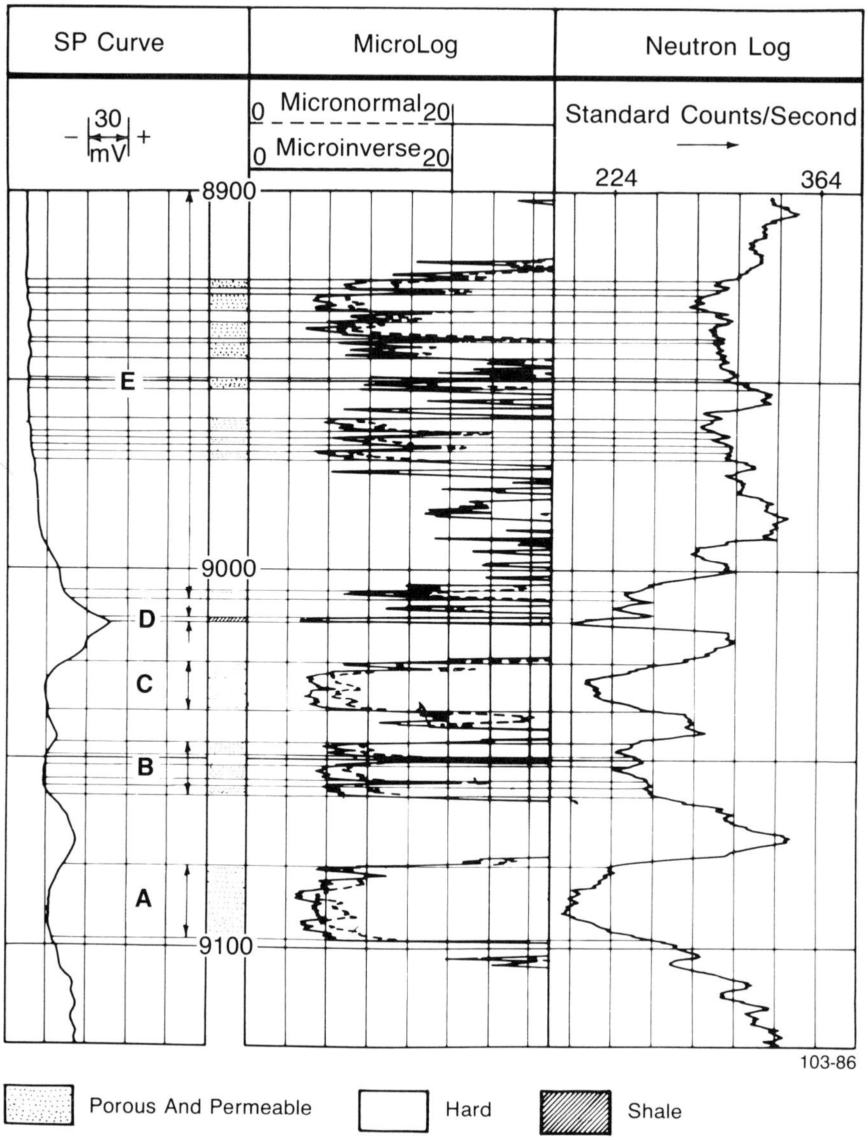


Fig. 45 Neutron log and MicroLog in hard formations.

# 10 Log Interpretation

## INTERPRETATION TECHNIQUES

Three interpretation techniques used with old logs are covered in this chapter. They are:

1. Ratio methods
   - Rocky Mountain method
   - $R_i$ method
   - $R_{xo}$ method
2. Shaly Sand method
3. Laterolog - $R_{xo}$ method

All these methods determine water saturation ($S_w$) based upon the Archie formula or variations thereof:

$$S_w^n = (F\ R_w)/R_t = R_0/R_t$$

- $n$ is usually 2 unless other local information is available.
- $R_w$ should be found from local knowledge of connate water salinity or determined from the SP curve; see Charts A-10 and A-12 in Appendix A.
- $R_0$ is the resistivity of the formation when $S_w$ = 100%. It can be found opposite a clean water-bearing zone.
- $R_t$ is determined from the E-log or Laterolog; see Chart B-7.
- $F$ is a function of porosity (Chart C-10) and can be determined from the short normal, the limestone curve, the MicroLog, the Microlaterolog, or the neutron curve.

The relationship between formation factor, $F$, and porosity, $\phi$ is given as:

$$F = \frac{a}{\phi^m} = \frac{R_0}{R_w}$$

For hard sands and limestones, $a = 1$ and $m = 2$ are good average values.

For soft, clean sands, $a = 0.81$ and $m = 2$ are good average values.

Local knowledge may offer more suitable $m$, $n$, and $a$ values.

Other porosity-$F$ relationships are: $F = R_i/R_z$ and $R_{xo}/R_{mf}$.

### Ratio Techniques

In resistivity-ratio methods it is assumed that the formation is divided into two distinct regions — a flushed zone and a noninvaded zone — both having the same $F$, but each containing water of a distinct resistivity ($R_{mf}$ or $R_z$ in the flushed zone and $R_w$ in the noninvaded zone). The bulk resistivities of the two zones must be available from logs, and there must be methods for determining the resistivity of the water in each zone.

Although all these methods have limitations because of the necessary assumptions, many have found wide use as wellsite techniques. They are sometimes the log analyst's only choice when interpreting old logs with no independent porosity data at hand.

Hydrocarbons create another difficulty. In this case, some knowledge or assumption of the value of flushed zone saturation ($S_{xo}$) is necessary.

#### Rocky Mountain Method

First introduced in 1949, the Rocky Mountain method permits, under certain conditions, the determination of water saturation and porosity when only $R_t$, $R_i$, and SP are known. This is because experience has shown that $S_i^2$ is approximately equal to $S_w$.

$R_i$ is obtained from the 16-in. normal after corrections (Chart B-2). $R_t$ is obtained from the 64-in. long normal or the 18-ft 8-in. lateral after correction (Chart B-2). The SP must be fully developed.

Water saturation can be found using Chart D-8, a graphical solution of the equation used to find water saturation. Chart D-10 can be used to find the porosity. This technique assumes that the short normal has the ability to measure porosity.

There are some limitations:

- Moderate to deep invasion is necessary. Hard-rock areas with low to medium porosity are essential.
- $R_{16''}/R_m$ must be greater than 10.
- Not applicable in salt muds.
- Not suitable in thin beds.

$R_i$ Method

In the Rocky Mountain method the term $F$ was eliminated from the saturation equation. If $F$ is known, the method can be revised so that $R_t$ is not required. This will be applicable in many instances where a MicroLog or Microlaterolog is available but the E-log is a cut-down version with either the long normal or the long lateral missing.

Suggested uses are:

- In thin resistive beds where the 16-in. normal approaches $R_i$ or can be corrected to $R_i$ by departure curves.
- When invasion has affected the lateral curve, use the 16-in. normal to get $R_i$.
- In fresh muds, where the Laterolog reading approaches $R_i$, this method is helpful in both thin resistive beds and in conductive beds.
- Use with salt muds in carbonates with very deep invasion, where the Laterolog may be reading closer to $R_i$ than to $R_t$.

In the $R_i$ method, the water saturation equation can be written

$$S_w = \frac{F\,R_z}{R_i}\ .$$

It can be shown that

$$S_w = \frac{R_i/R_t}{R_z/R_w}\ .$$

$R_z$ can be obtained from the equation

$$\frac{1}{R_z} = \frac{z}{R_w} + \frac{1-z}{R_{mf}},$$

where $z$ is the percent of formation water in the invaded zone ($R_i$). The value of $z$ can be varied to suit experience:

| | Porosity | $z$ | Invasion |
|---|---|---|---|
| For Gulf Coast-type soft sands | 30-35 | 0.15 | Shallow |
| | 25-30 | 0.10 | Shallow |
| For midcontinent-type sands | 15-25 | 0.08 | Moderate |
| For carbonates and tight sands | 10-15 | 0.05 | Deep |
| | 5-10 | 0.03 | Deep |

This technique also has limitations:

- Saturation can be studied in clean formations only.
- $R_{16''}/R_m$ must be greater than 10.

The $R_i$ method is not as accurate as other methods that use $R_t$ to determine saturation. In particular, errors on the formation factor will greatly affect the answer. A reliable value of $R_t$ is always best for determining $S_w$.

$R_{xo}$ Ratio Method

If a MicroLog or a Microlaterolog is available, $R_{xo}$ can be determined and the accuracy of the interpretation will be greatly improved. Other necessary inputs are $R_t$ from the E-log and $R_{mf}/R_w$, either measured directly or obtained from the SP curve. No knowledge of porosity or formation factor is needed. An average residual oil saturation (ROS) correction is built into the model based upon an empirical relationship between $S_w$ and $S_{xo}$.

Chart D-4 provides a graphical solution to the equation used to find water saturation.

Limitations of this technique:

- If the E-log is used, not suitable for thin beds.
- Poor results in large or caved holes.

**Shaly Sand Method**

As we have seen, any method that gives correct values for $R_i$, $R_{xo}$, $R_t$, $R_{mf}$, and $R_w$ can be used to determine water saturation when ROS is known or can be surmised.

The practical method of interpretation in shaly formations is based upon the observations made on field logs: in 100% water-saturated shaly sands, the pseudostatic SP (PSP) is given by:

$$\text{PSP} = -K \log R_{xo}/R_0$$

This equation was also verified in the laboratory and is mathematically sound for a formation consisting of beds of thin, laminated, clean sand and pure shale.

In a shaly sand, knowledge of $R_{xo}$, $R_t$, PSP, and static SP (SSP) is required. The ROS should be estimated. The SSP can be determined if some practically clean sands are found reasonably close to the shaly formation considered. The PSP is given by the deflection of the SP curve at the level of the bed.

Chart D-4 is also used to determine water saturation in shaly formations. Chart D-6 is a companion chart that provides a method of determining the effective porosity in shaly sands. It uses much the same data as before but also requires the alpha factor, which is the ratio of PSP to SSP. A detailed explanation of how to use each chart is provided in Appendix A.

Limitations of this method:

- Applicable only when the proportion of shale is not too great.
- Not applicable in salt muds.
- If the E-log is involved, not suitable in thin beds.
- Poor results in large or caved holes.

**Laterolog - $R_{xo}$ Method**

In hard formations drilled with very salty muds, the standard logging program usually consists of the Laterolog (either 3 or 7), the Microlaterolog, and the gamma ray and neutron logs.

The Laterolog, after correction for invasion, will provide a good value of $R_t$. The Microlaterolog will read $R_{xo}$

directly except when the formation is vuggy or fractured or when $t_{mc}$ > ½ in. In such cases, the neutron log is preferred as a source of porosity.

Chart D-15 provides a means of computing $S_w$ when the Laterolog is used for the measurement of $R_t$ and the Microlaterolog is used for the measurement of $R_{xo}$.

Limitations of this method:

- Not suitable in high-porosity or unconsolidated formations.
- ROS is often difficult to estimate in vuggy, fractured formations.
- Poor results in large or caved holes.

In most cases, clean oil or gas production demands a water saturation less than 50% in sands and sandstones. The critical value of saturation seems to be much less for clean oil or gas production in limestones and dolomites. In sandstones, a porosity of less than 8% is usually too low for commercial production.

## INTERPRETATION EXAMPLES

### Log Example 1

The following pages show hand-drawn E-logs and MicroLogs that illustrate all the different idiosyncracies of the old surveys. They are intended to be a reference guide for prospecting for oil and gas using old logs.

C.K. Ruddick drew these original "type logs" in 1956. His pictorials have been used by log analysts throughout the years as a training aid and as a minirefresher course. They cover all the responses associated with old E-logs and the MicroLog.

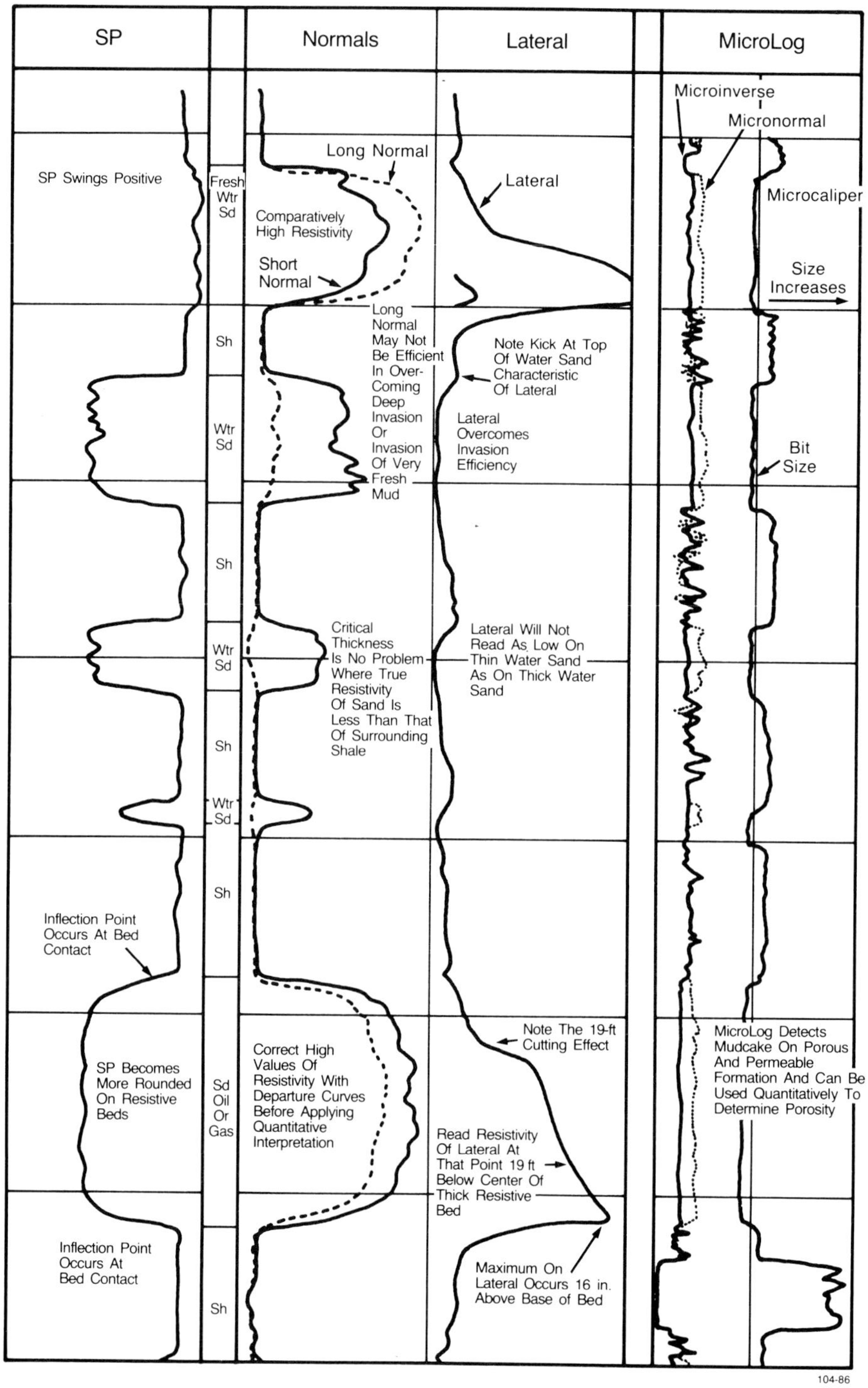


Fig. 46 Log example 1.

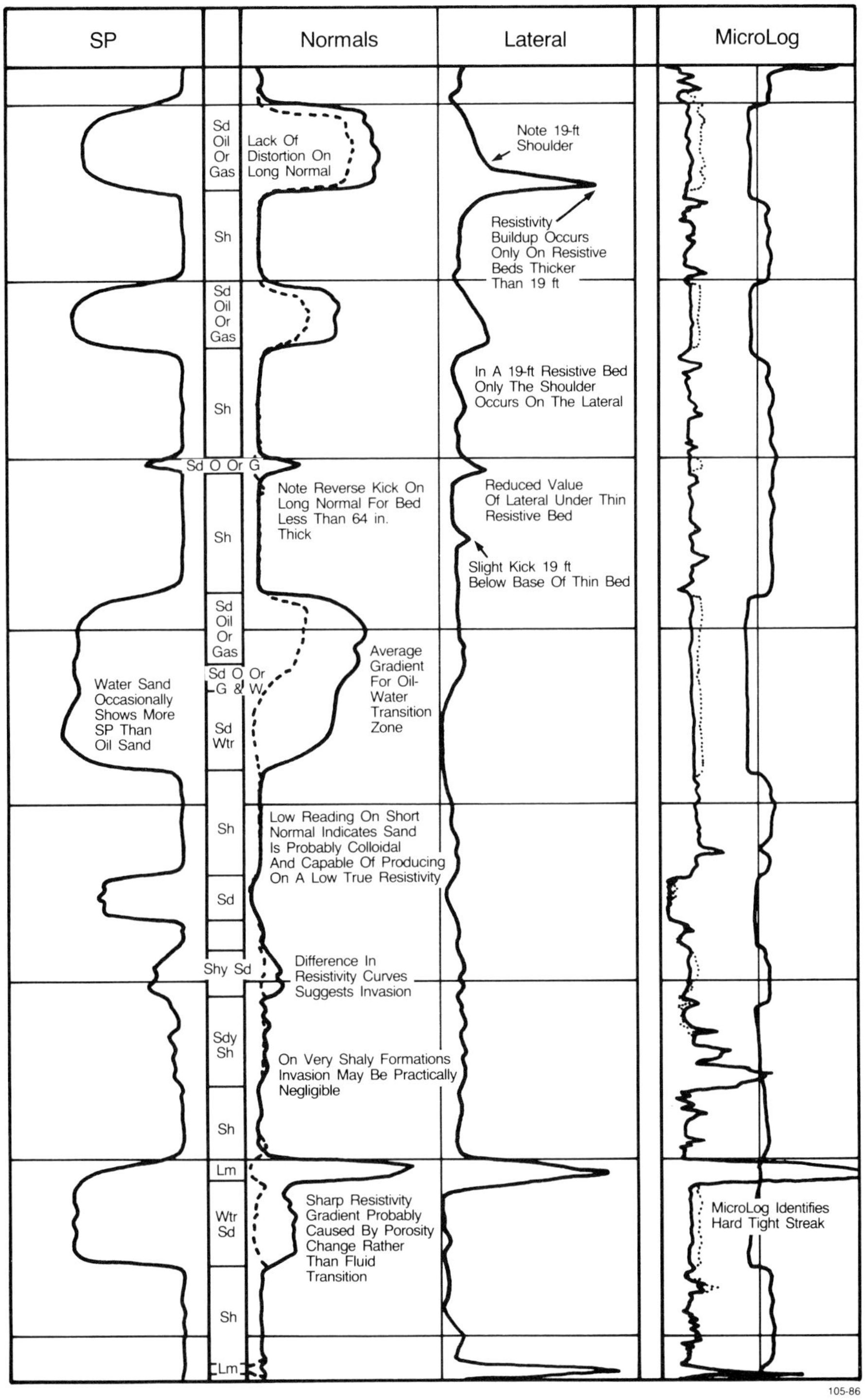


Fig. 47 Log example 1.

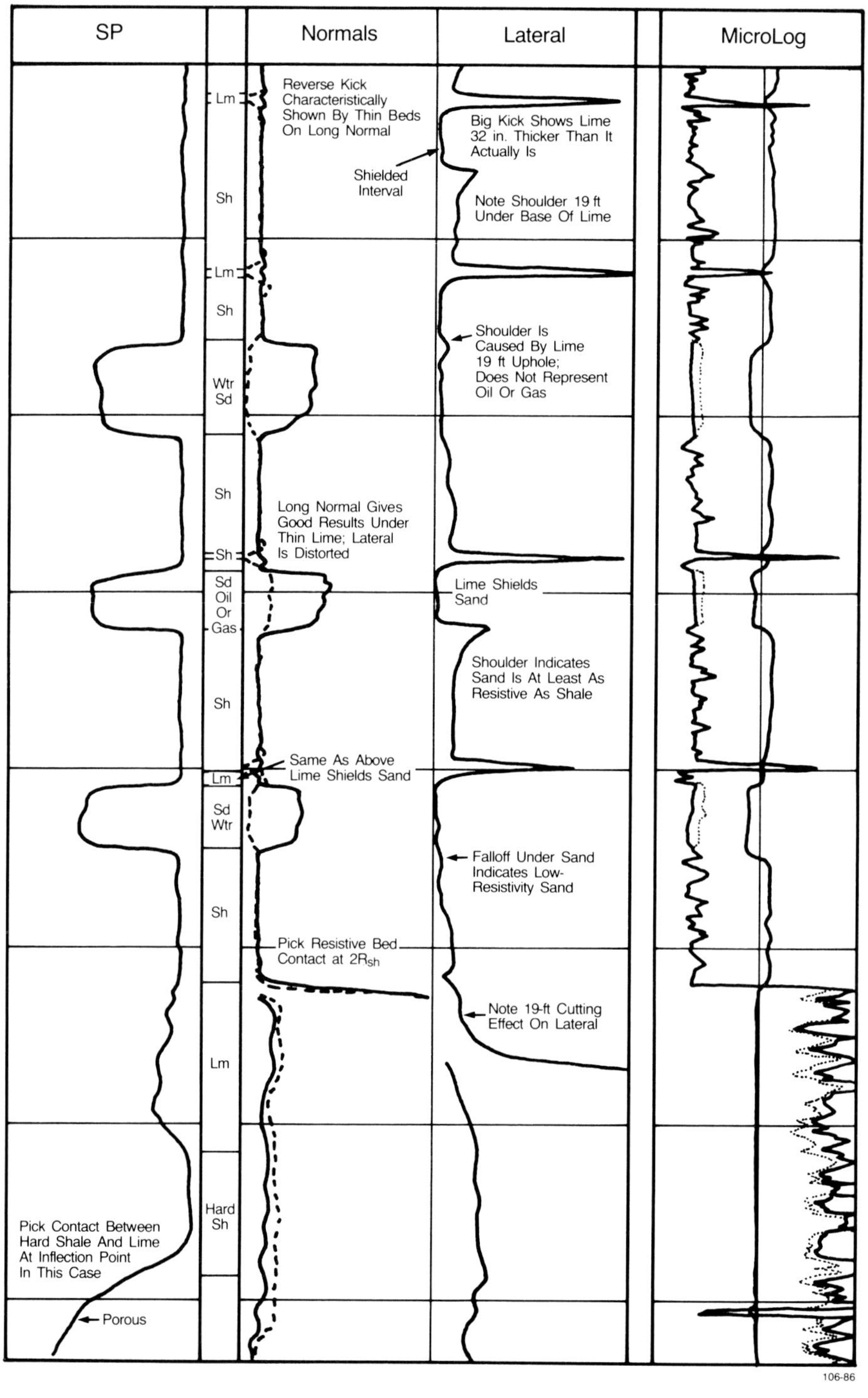


Fig. 48 Log example 1.

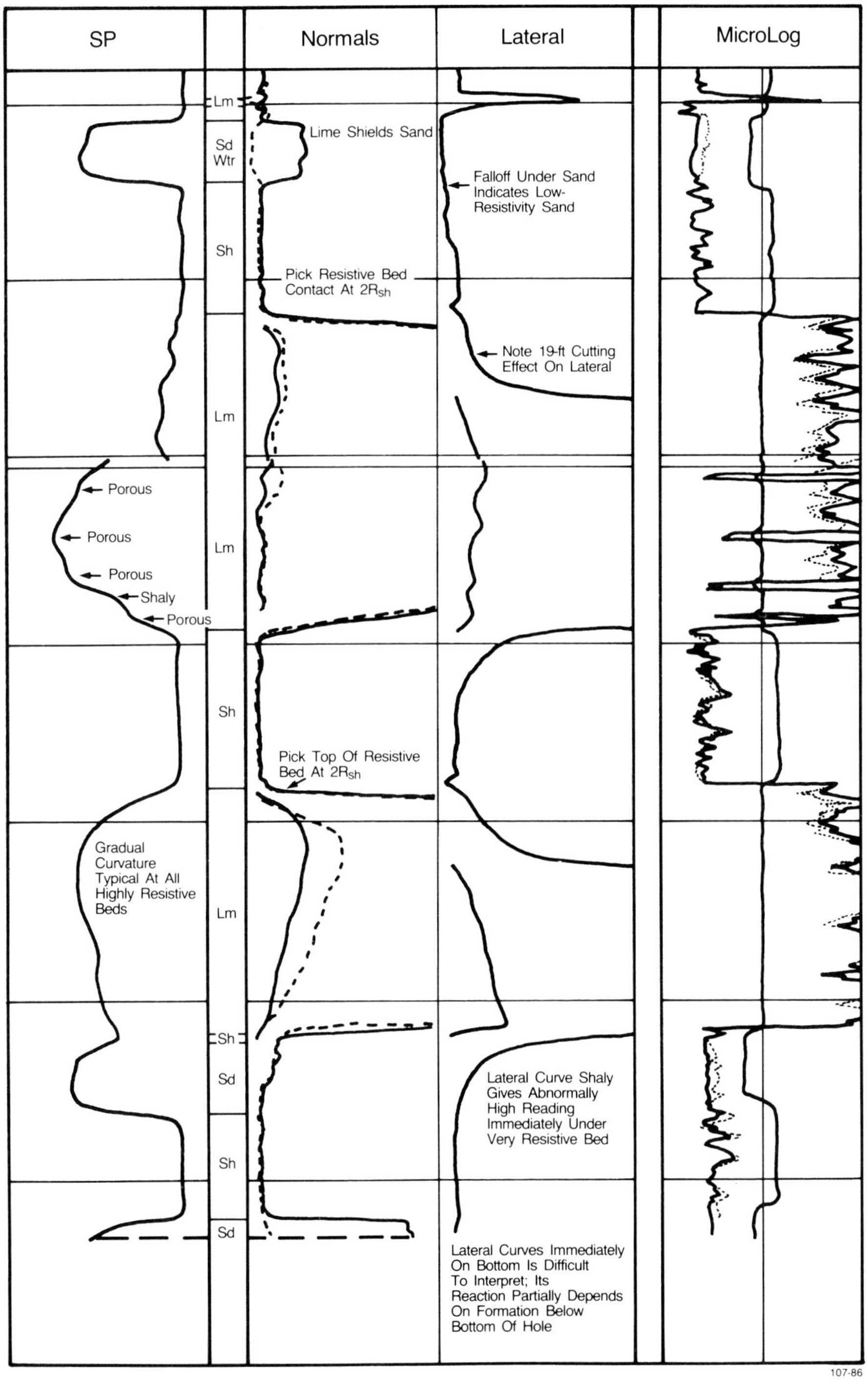


Fig. 49 Log example 1.

**Log Example 2**

**Background:** Water well, Brazoria County, Texas - 1954.

**Logs:** Electrical survey only.

**Given:** $R_m = 1$ at 77° F.

**Instructions:** How many clean (nonshaly) water sands are on this log?

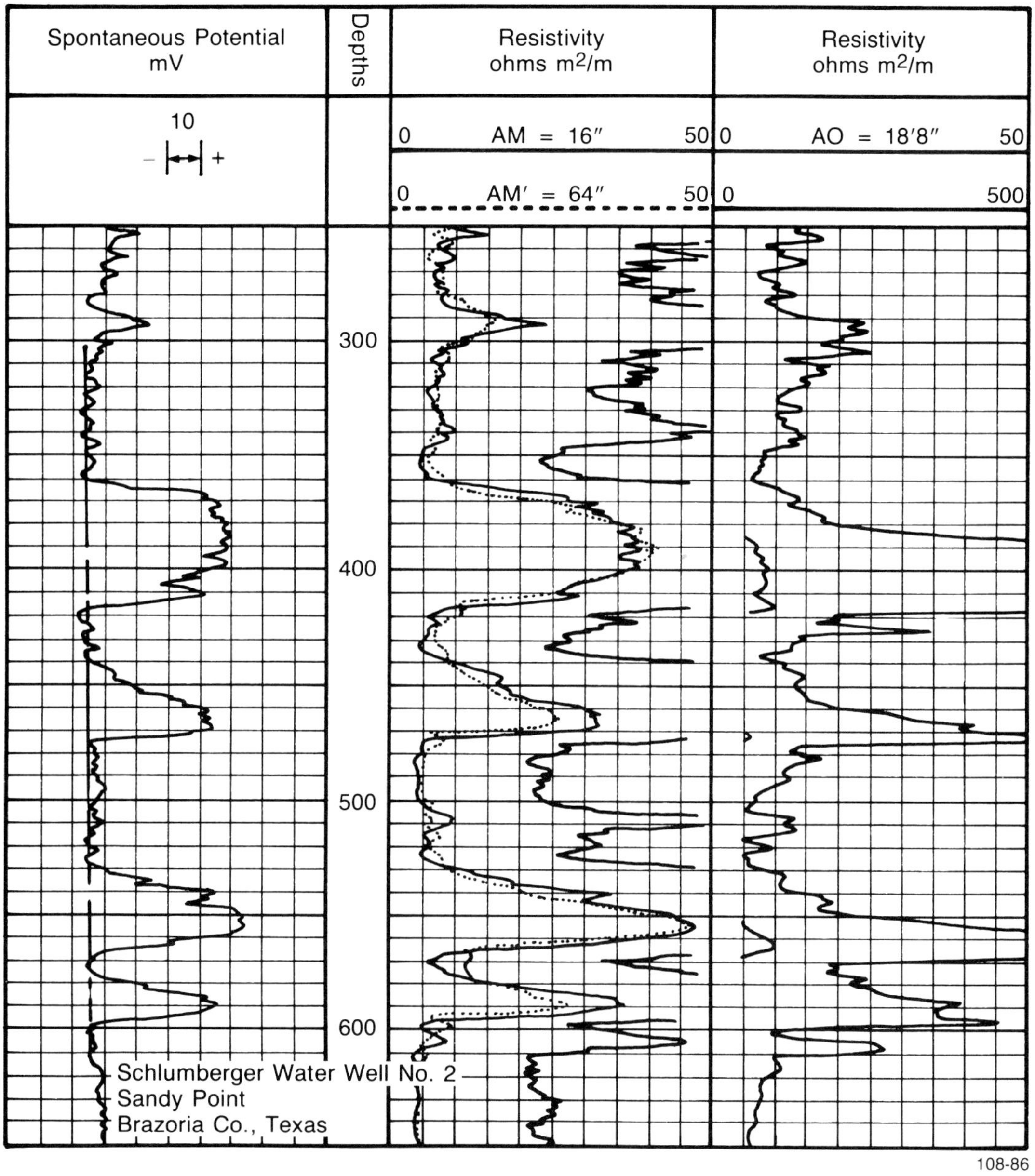


Fig. 50 Log example 2.

### *Log Example 2 Answers*

*There are four clean freshwater sands on this log. The SP curve is reversed (positive) because of the fresh water in the aquifer. Wells drilled for fresh water are evaluated for the quality and quantity of water that may be obtained from various strata. The SP curve is used to determine quality and the resistivity measurement provides the means for determining relative productivities of freshwater sands. For a complete discussion of this technique, refer to* Interpretation of Electric Logs in Freshwater Wells in Unconsolidated Formations *by R.P. Alger, available from Schlumberger Educational Services.*

### Log Example 3

**Background:** Grayson County, Texas - 1956.

**Logs:** Halliburton E-log only.

**Given:** Three resistivity curves plus an SP.

**Instructions:** Read the remarks section of the heading.

What type resistivity curve is a 2Z18 in.?

What type resistivity curve is a 3iZ16 ft?

What type resistivity curve is a 3iZ9 ft?

At what depth on the log can the curves be identified?

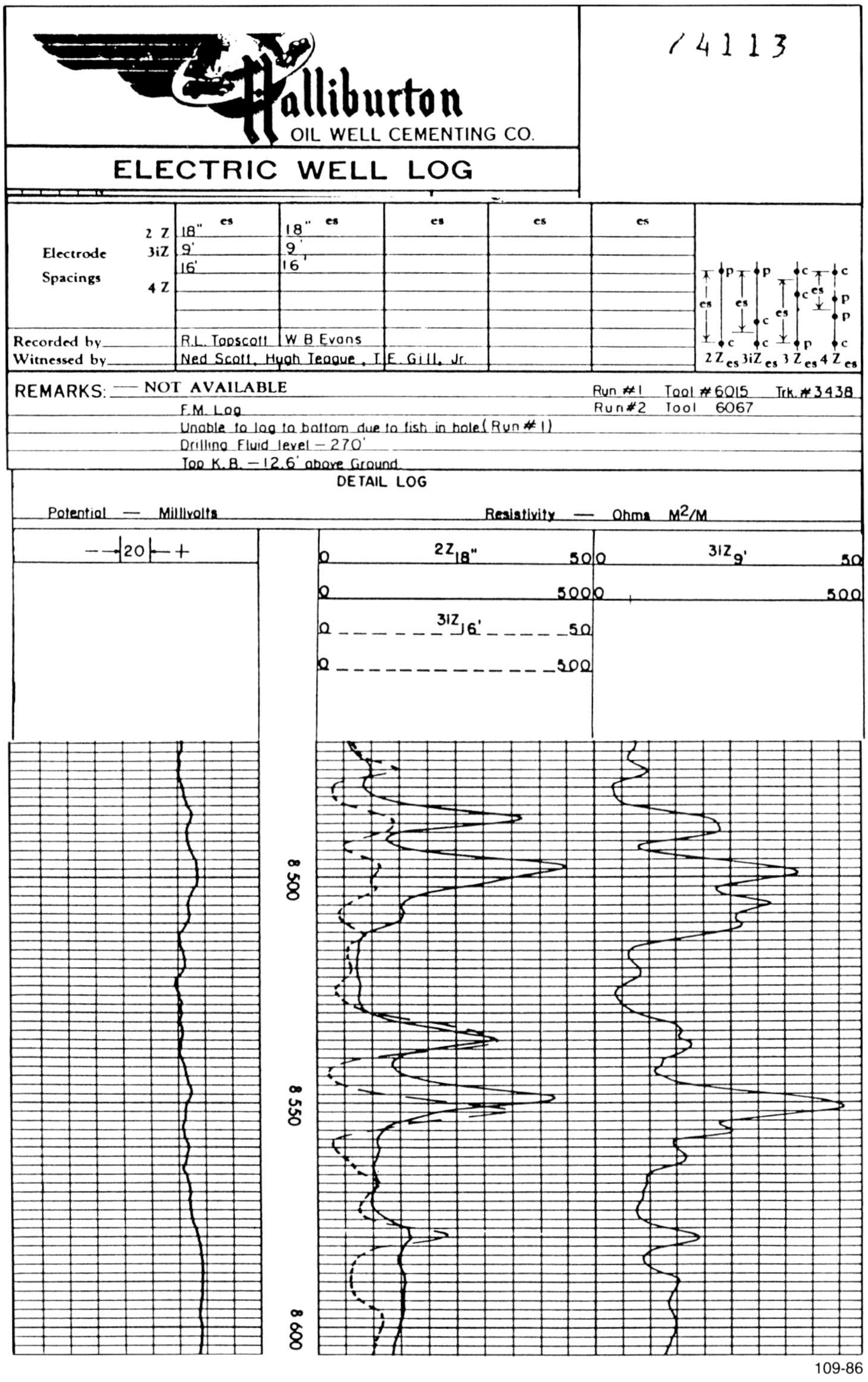


Fig. 51 Log example 3.

### *Log Example 3 Answers*

*A 2Z18 in. is an 18-in. short normal curve. A 3iZ16 ft is a 16-ft lateral curve. A 3iZ9 ft is a 9-ft lateral curve. The Halliburton electric log was sometimes referred to as the "jeep." It featured a frequency modulation (FM) system. Current electrodes were termed "c" and measure electrodes were designated as "p." Both lateral spacings can be identified at the hard streak at 8580 ft. Both the 9-ft and 16-ft blind zones are evident here.*

### Log Example 4

Background: In the Texas and Louisiana Gulf Coast areas, many productive zones are very thin, less than 5 ft thick. Before the short normal curve can be used to determine true resistivity, it must be corrected for bed thickness. The SP curve must also be corrected for thin-bed effect before it can be used to determine $R_w$.

These are Miocene gas sands in south Louisiana. The overall interval is 3845-3865 ft.

Logs: 1950 E-log only.

Given: $R_w$ = 0.06 ohm-m at 3900 ft.

$R_{16''}$ = 5 ohm-m.

$R_m$ = 1.4 ohm-m at BHT.

Porosity = 30% from sidewall cores.

Hole size = 8 in.

Instructions: At 3850-3853 ft correct the short normal curve for thin-bed effect using Chart B-10.

Correct the SP curve for thin-bed effect using Chart A-8.

Calculate $S_w$.

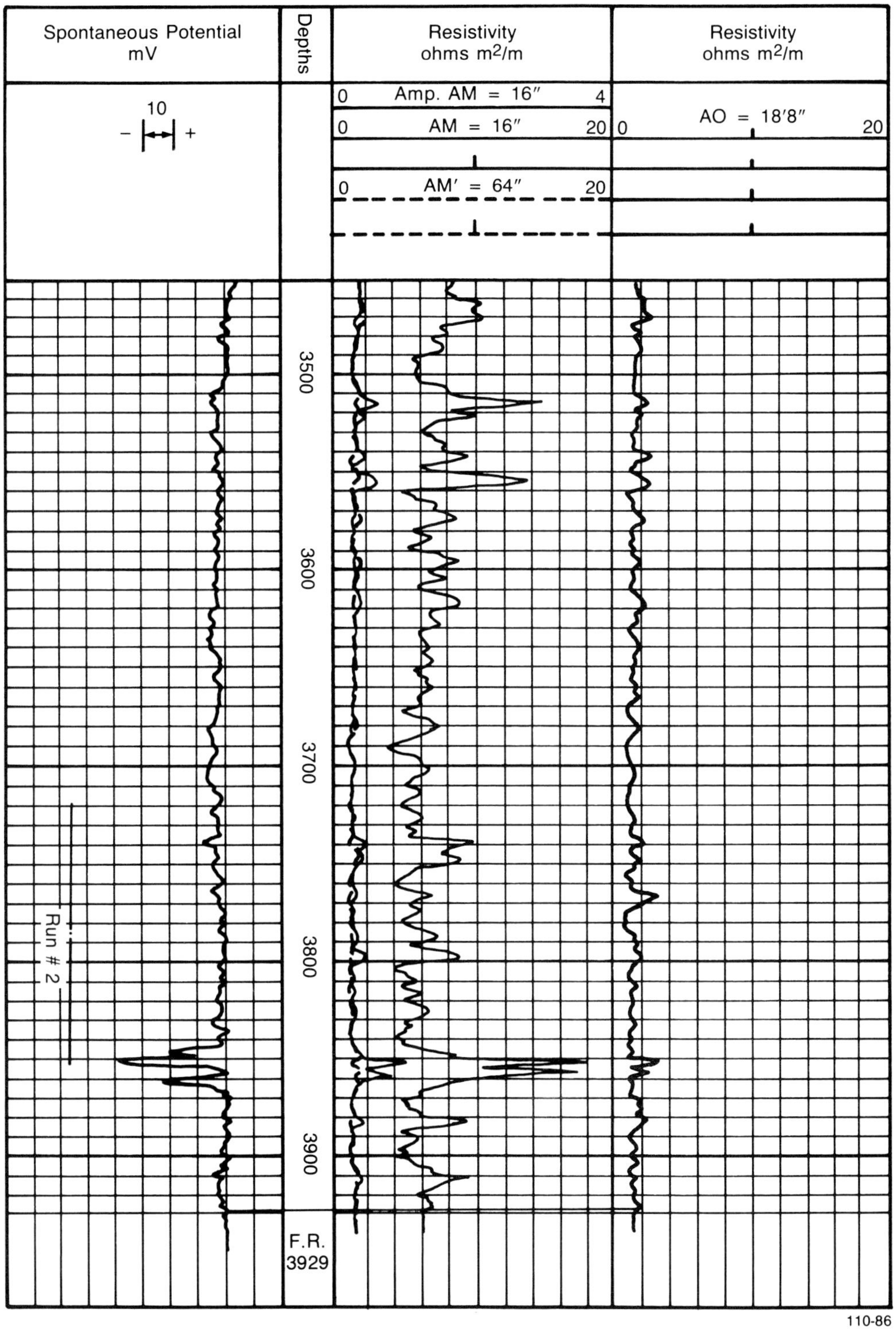


Fig. 52 Log example 4.

### *Log Example 4 Answers*

*$R_{16''}/R_m$ = 5/1.4 = 3.6. Using Chart B-10 and 8-in. hole, $R_{16''}/R_m$ corrected = 7. $R_{16''}$ = 7 x 1.4 = 9.8 ohm-m = $R_t$ (see Chart B-8). For porosity = 30%, F = 8. $F = R_0/R_w$ and $R_0 = F R_w$ = 8 x 0.06 = 0.48. $S_w = \sqrt{R_0/R_t} = \sqrt{0.48/9.8}$ = 22%. SP correction (Chart A-8): $R_i$ = 10; $R_m$ = 1.4; SP = 38 mV. $R_i/R_m$ = 7; for a 3-ft bed, the correction factor is 1.67. SP corrected = 38 x 1.67 = 63 mV.*

### Log Example 5

**Background:** Miocene - offshore Louisiana.

Oil sand at 6962-6275 ft.

**Logs:** 1952 E-log and MicroLog.

**Given:** See data on log example.

Use $K$ = 80.

Assume ROS = 20%.

**Instructions:** Using the shaly sand interpretation technique, calculate porosity and $S_w$ at 6962-6975 ft.

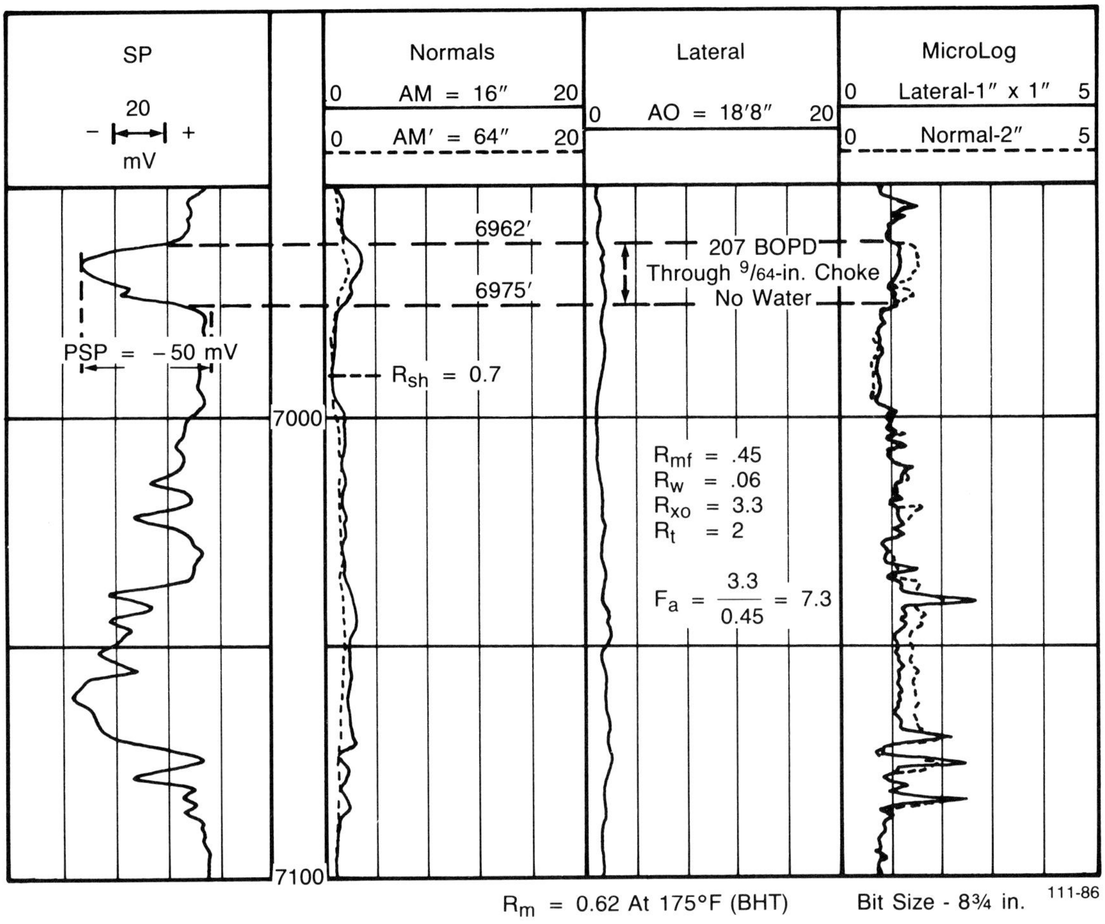


Fig. 53 Log example 5.

### *Log Example 5 Answers*

*$R_{xo}/R_t$ = 1.65; $R_{mf}/R_w$ = 7.5; PSP = 50 mV; SSP = 70 mV; PSP/SSP = 0.71. Using Chart D-4, $S_{wa}$ = 54% and $S_w$ = 42%. Using Chart D-6, $R_{xo}/R_{mf}$ = 7.3 and effective porosity = 29%. With no correction for shale, the apparent porosity would have been 39%. Hundreds of articles on shaly sand evaluations have been written over the years. Many different methods have been proposed with old E-logs, but only one technique has given satisfactory results. This example demonstrates that technique. See Chart D-3 through D-6 for a detailed explanation of this shaly sand interpretation technique.*

### Log Example 6

**Background:** Permian sand - Coleman County, Texas.

Many log libraries sold a cut-down version of the original E-log. Track 3 was frequently cut off and destroyed. Depending upon the areas, the cut-off portion contained either the long normal (as in this example) or the long lateral resistivity curves. Obviously, with one important curve missing, log evaluation was severely hampered. Log Example 6 deals with this problem.

**Logs:** 1959 E-log only.

**Given:** $R_m$ = 1.20 ohm-m at 110° F (2900 ft).

$R_w$ = 0.04 ohm-m at 110° F.

Standard spacings.

Moderate invasion.

**Instructions:** Interpret this interval: 2940-2955 ft.

Determine porosity from the short normal curve.

Calculate $S_w$ using:

- The Archie equation.
- The Rocky Mountain method.

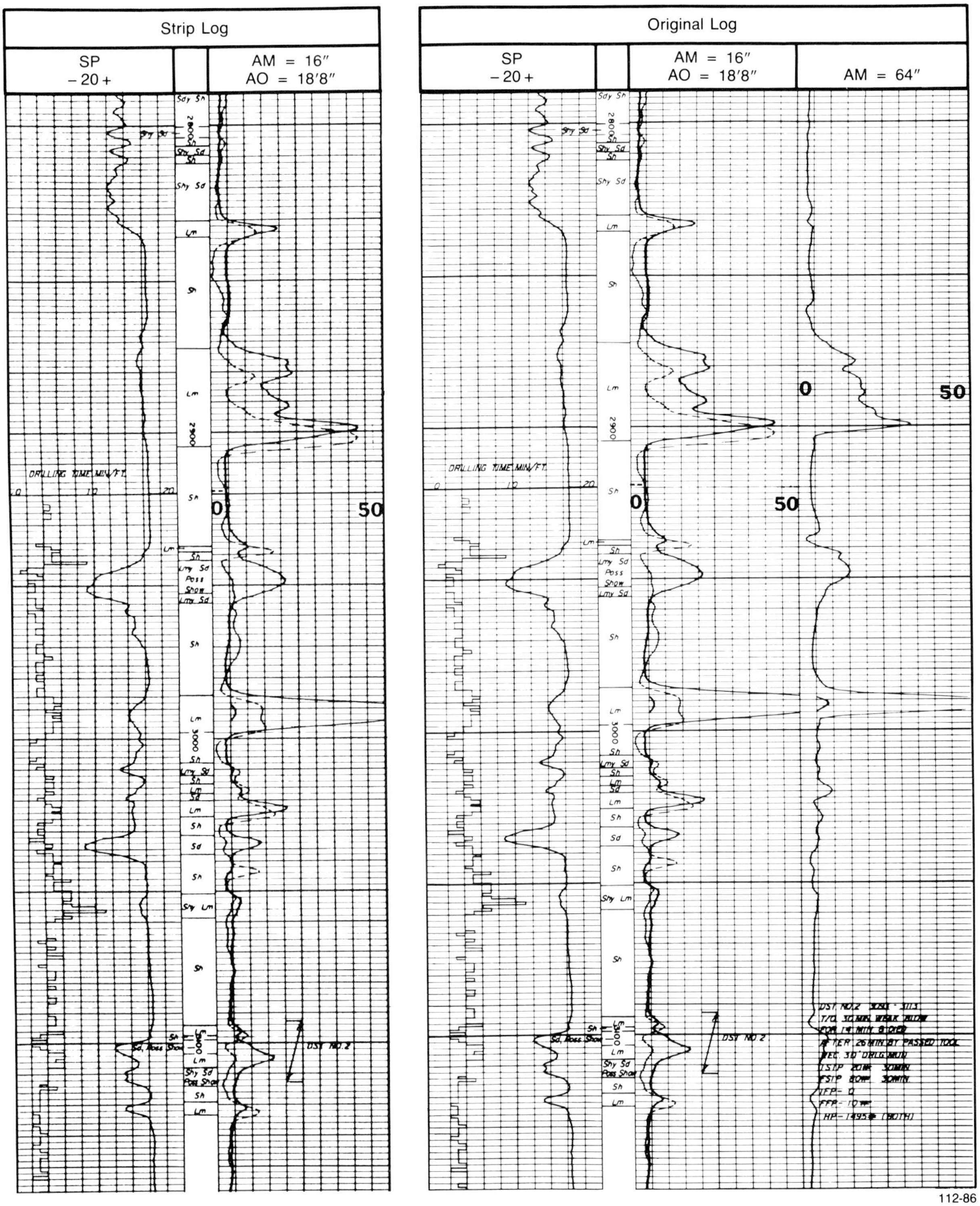


Fig. 54 Log example 6.

### *Log Example 6 Answers*

*Porosity from the short normal curve:* $R_{16''} = 22$ *ohm-m;* $R_{64''} = 15$ *ohm-m; Lateral = blinded;* $R_{mf} = 0.95$ *ohm-m at 110° F;* $z = 0.08$ *(moderate invasion). Calculate* $R_z$*:* $1/R_z = 0.08/0.04 + 0.92/0.95 = 2.97$*;* $R_z = 0.34$*; Calculate* $F_a$*:* $F_a = R_i/R_z = 22/0.34 = 65$*. Assuming 20% ROS,* $F = 65\,(1 - 0.20)^2 = 42$ *and porosity = 15%. Water saturation: Archie equation:* $S_w = \sqrt{42 \times 0.04/15} = 33\%$*. Rocky Mountain method:* $R_i/R_t = 22/15 = 1.47$*; SP = 76 mV;* $S_w = 31\%$*; porosity = 15%. This well was re-entered and produced 800,000 cubic feet of gas per day from this interval on a 16/64-in. choke.*

### Log Example 7

**Background:** Lower cretaceous limestone. Cherokee County, Texas.

Prospecting with old E-logs and MicroLogs in carbonates is possible if the porosity is greater than 8%. The Pettit limestone in east Texas is typical of many carbonates found in the basins of North America.

**Logs:** 1957 E-log and MicroLog.

**Given:** $R_m = 0.33$ ohm-m at 185° F.

$R_w = 0.02$ ohm-m at 8500 ft.

Standard electrode spacings.

**Instructions:** Interpret this interval: 8460-8480 ft.

Calculate porosity from the short normal.

Determine porosity from the empirical MicroLog chart.

Calculate $S_w$ from a ratio method.

Calculate $S_w$ using the Archie equation.

What is the average mudcake thickness over this zone?

What is the net pay thickness?

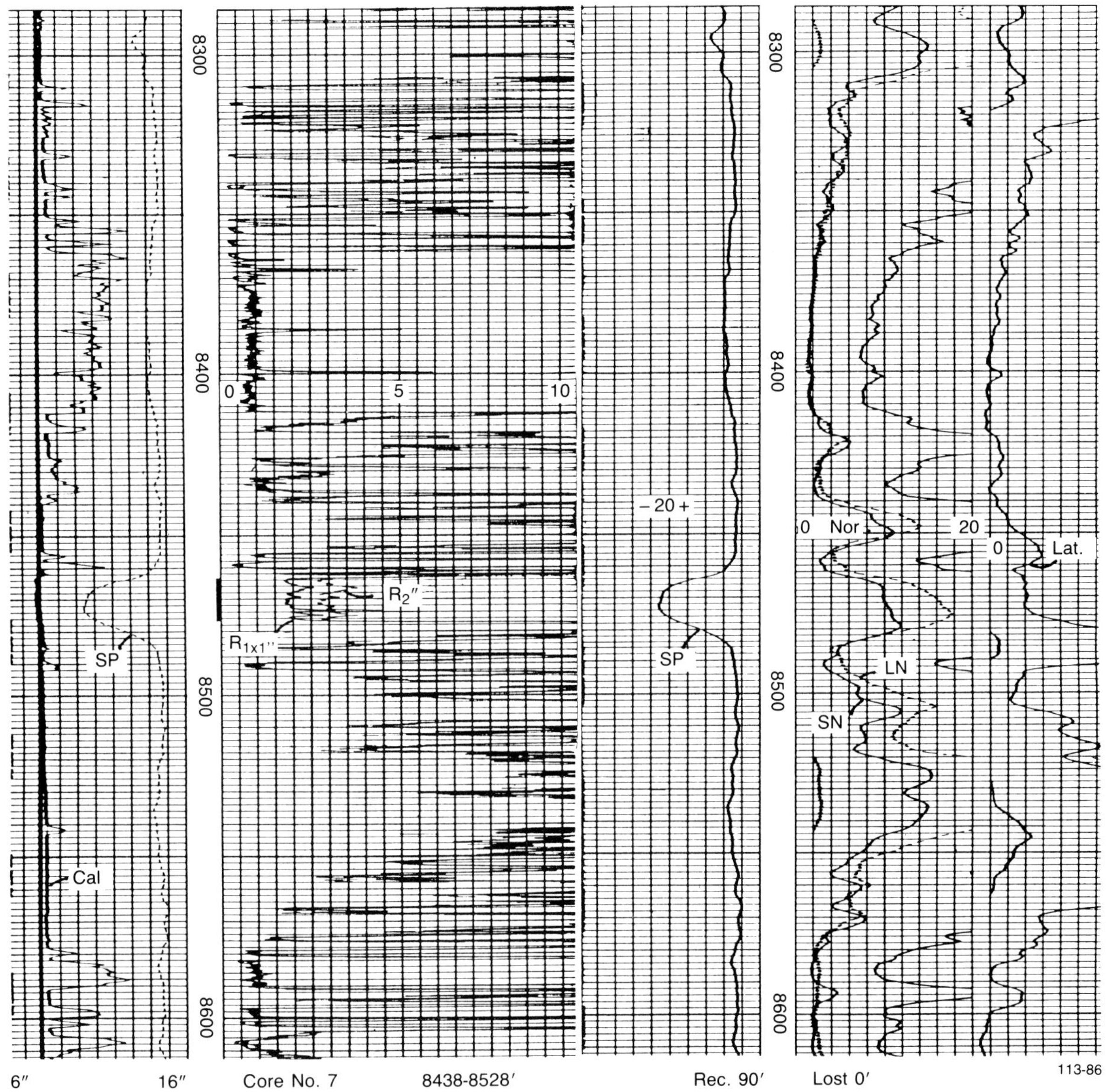


| | |
|---|---|
| 2 ft | Black Calcareous Shale |
| 11 ft | Nonporous, Dark Gray Finely Crystalline Shaly Pseudooolitic Limestone With Thin Streaks Black Calcareous Shale, No Show |
| 9½ ft | Black Calcareous Shale With Thin Beds Nonporous, Dark Gray Dense Shaly Limestone, No Show |
| 15½ ft | Slightly Porous Grayish Brown Oolitic Slightly Fossiliferous Limestone With Faint Gas Odor And Taste With Occasional Thin Streaks Being Very Slightly Porous And Nonporous, Top 6 in. Being Shaly Nonporous |
| 6 ft | Nonporous Dark Gray Finely Crystalline Shaly Slightly Fossiliferous Oolitic Limestone With Irregular Thin Streaks Black Calcareous Shale, No Show |
| 17 ft | Black Calcareous Shale With Thin Bed Dark Gray Dense Shaly Limestone, No Show |
| 20 ft | Nonporous Dark Gray Finely Crystalline Shaly Slightly Oolitic And Fossiliferous Limestone With Thin Streaks And Numerous Thin Beds Black Calcareous Shale |

Fig. 55 Log example 7.

### *Log Example 7 Answers*

*Porosity from the short normal = 12%; Porosity from the MicroLog chart = 12-13%; $S_w$ from a ratio method = 19% (Chart D-4) and 11% (Chart D-8); $S_w$ from the Archie equation = 26-27%. Mudcake is thin - less than 1/8 in. Net pay thickness = 13 ft. This well produced 7.1 billion cubic feet of gas.*

### Log Example 8

**Background:** Upper cretaceous sand - east Texas.

**Logs:** E-log and MicroLog.

**Given:** $R_m$ = 1.70 ohm-m at 135° F (TD).
$R_w$ = 0.05 ohm-m at 135° F.
Standard spacings.

**Instructions:** Determine porosity and $S_w$ at 4510-4520 ft.
What is the mudcake thickness at 4515 ft?
What is the temperature gradient in this well?
Should pipe be set on this well?

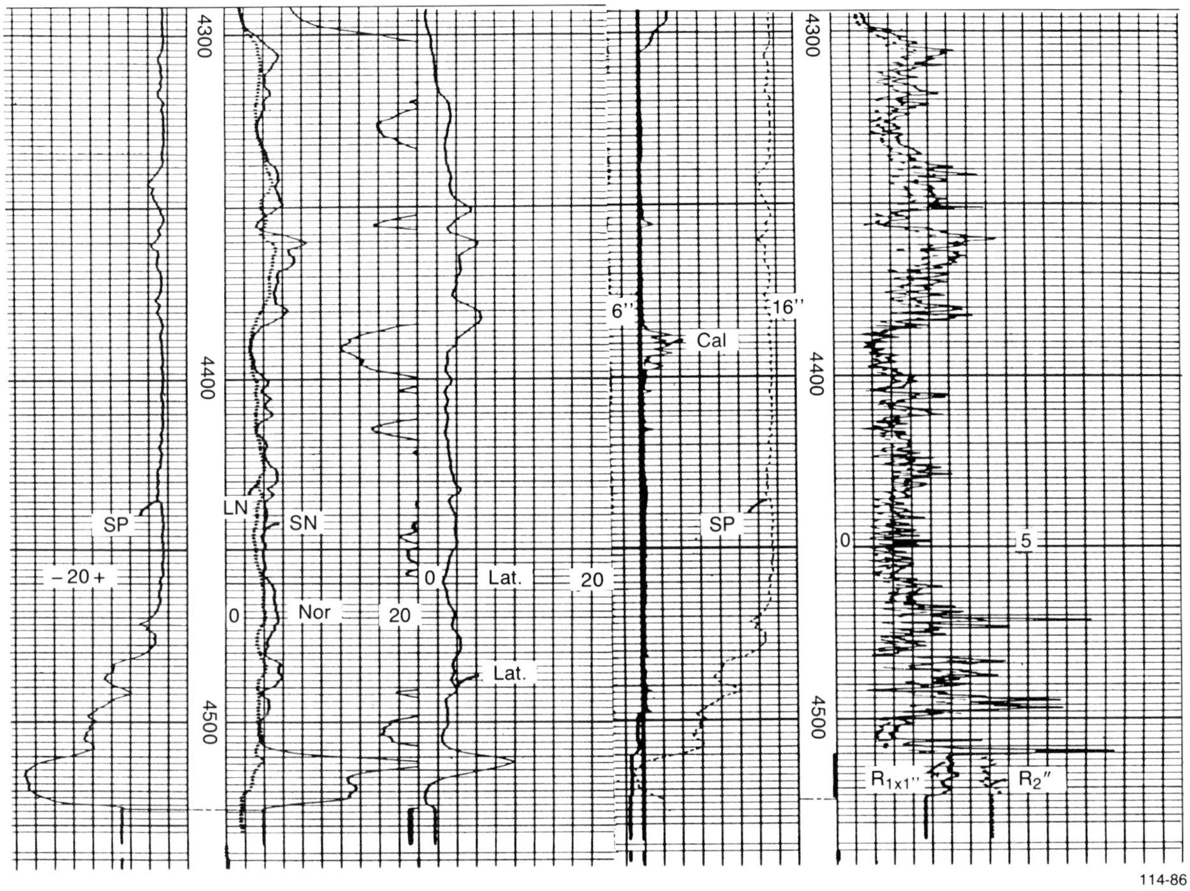


Core No. 1 4495′-4529′ Rec. 25½″ Lost 8½″

| | |
|---|---|
| 9 ft | Light Gray Shaly Chalk |
| 3 ft | Nonporous Very Sandy Shaly Fossiliferous Chalk With Few Streaks Asphaltic Staining |
| 13½ ft | Porous Fine To Medium Grain Sandstone With Light Brown Oil Stain And Odor |

Fig. 56 Log example 8.

### *Log Example 8 Answers*

*At 4510-4520 ft porosity = 28.5% (from empirical MicroLog chart). $S_w$ = 40-49% from the Archie equation. Mudcake is about 0.4 in. thick. Temperature gradient is about 1.25° F/100 ft. This sand appears to be in the transition zone. Generally, when logs and cores both have shows, one feels confident in deciding to set pipe. However, in this case the zone has a "wet-looking" resistivity profile and it should be tested with a DST or wireline formation tester before setting pipe.*

### Log Example 9

**Background:** Deep Wilcox sands - Colorado County, Texas.

**Logs:** E-log and MicroLog.

**Given:** See data on bottom of log.

**Instructions:** Use "eyeball" interpretation.
Should pipe be set on this well?

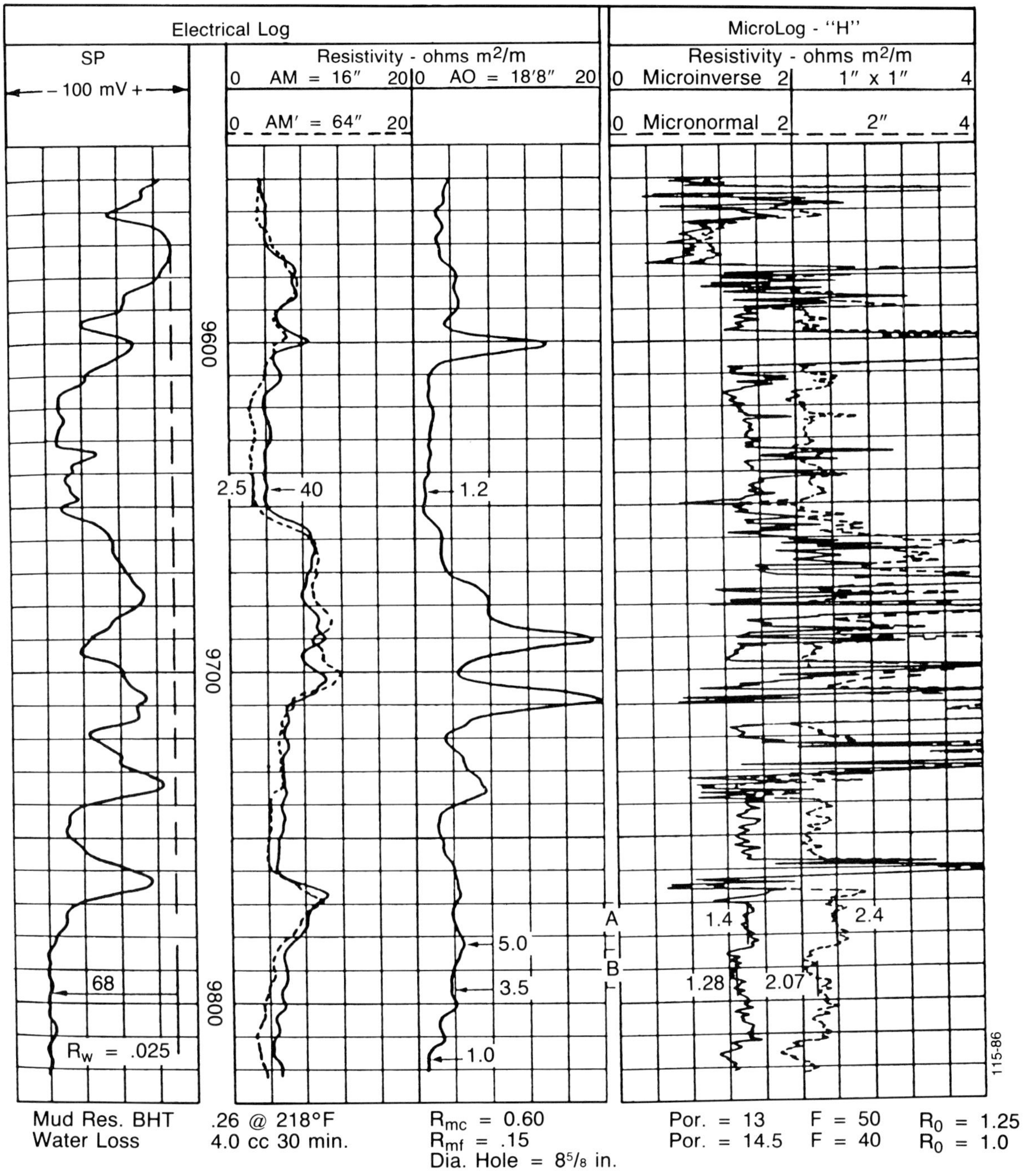


Fig. 57 Log example 9.

### *Log Example 9 Answers*

*Yes. There will likely be multiple pay zones. Some of them are: 9590-9597 ft, 9692-9698 ft, 9716-9722 ft, 9767-9780 ft. Below 9780 is transitional.*

### Log Example 10

**Background:** Pennsylvanian limestones. Runnels County, Texas.

**Logs:** 1954 E-log and MicroLog.

**Given:** $R_m$ = 0.90 ohm-m at 110° F at 3700 ft.

$R_w$ = 0.04 ohm-m at 110° F at 3700 ft.

$R_{mf}$ = 0.70 at 110° F (measured).

Hole diameter = 8 in.

Standard spacings.

**Instructions:** Compute $S_w$ and $R_z$ in both zones.

Upper zone: 3528-3534 ft.

Lower zone: 3590-3599 ft.

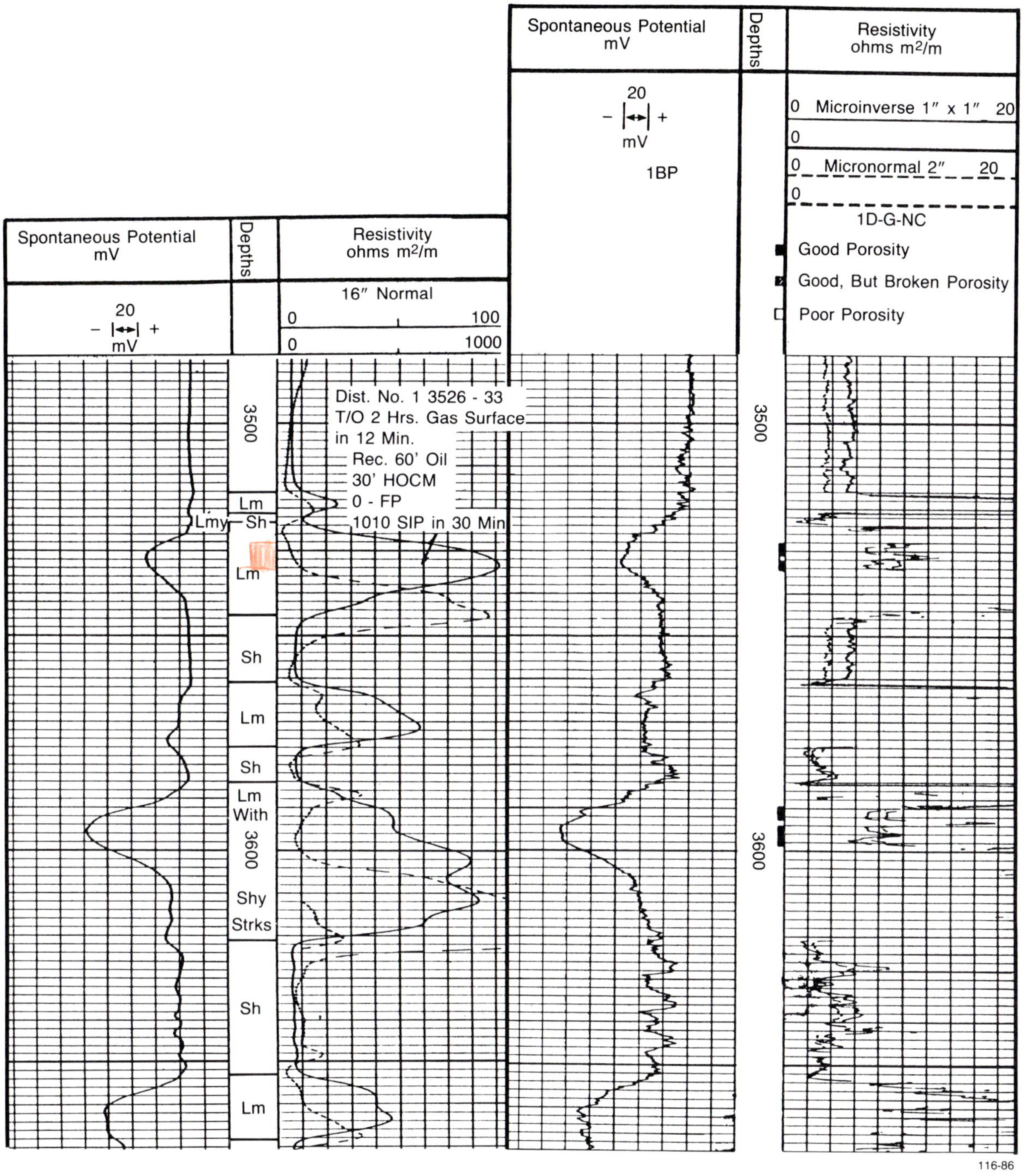


Fig. 58 Log example 10.

### *Log Example 10 Answers*

*In this example the short normal is the only useful resistivity curve. The long normal has been cut away by the log library and the lateral is blinded by tight lime streaks above the beds. When this occurs, water saturation can be estimated by the $R_i$ method where $S_w = F\,R_z/R_i$. To calculate $R_z$ (moderate invasion $z = 0.08$): $1/R_z = z/R_w + (1 - z)/R_{mf} = 0.08/0.04 + 0.92/0.7 = 3.3$; $R_z = 0.30$. Upper zone (3528-3534): Porosity = 9.5 and F = 110 (from MicroLog Chart C-2); $R_{16''} = 92$ ohm-m corrected by Chart B-2 (d = 8 in.) = 126 ohm-m. Using the $R_i$ method, $S_w = F\,R_z/R_i = 110 \times 0.30/126 = 0.26$. This would be hydrocarbon production. Lower zone (3590-3599): Porosity = 10 and F = 100 (from MicroLog Chart C-2); $R_{16''} = 48$ ohm-m corrected by Chart B-8 (d = 8 in.) = 48 ohm-m; $S_w = 100 \times 0.30/48 = 0.63$. This would be water production. These are ideal conditions for using the $R_i$ method.*

### Log Example 11

**Background:** Lower Pennsylvanian sand - Grayson County, Texas.

**Logs:** 1955 E-log and MicroLog.

**Given:** $R_m$ = 1.0 ohm-m at 8600 ft.

$R_w$ = 0.08 ohm-m at 8600 ft.

$R_{mf}$ = 0.66 ohm-m at 8600 ft.

Standard E-log curves.

**Instructions:** Compute porosity and $S_w$ at the interval 8618-8647 ft.

What is the net pay footage in this interval?

Why isn't the SP developed?

The DST failed. Should pipe be set?

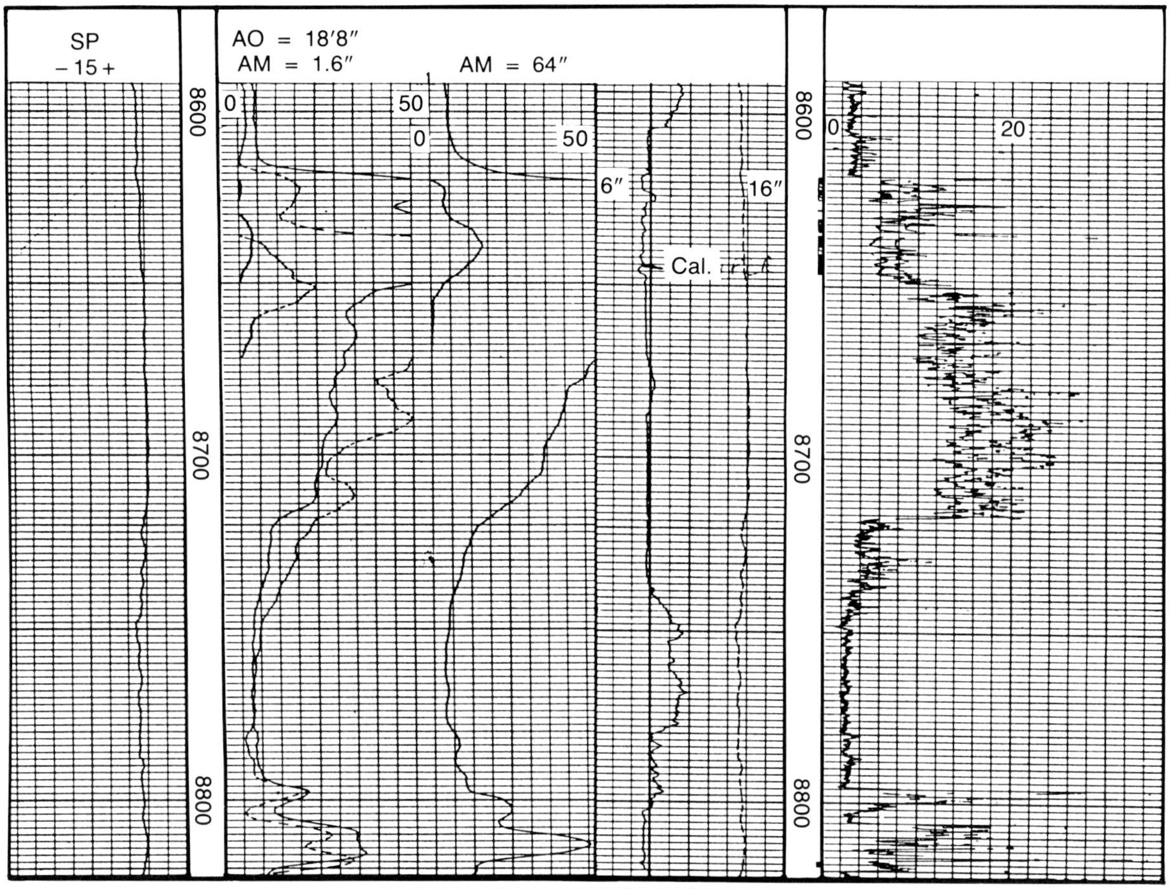


117-86

**SCHLUMBERGER WELL SURVEYING CORPORATION**
HOUSTON, TEXAS

QUANTITATIVE INTERPRETATION

Gentlemen

Tabulated below these comments are some average values of porosity and water saturation which we have calculated from the logs. Whenever sufficient information is obtainable from the logs, we indicate the possible production in the right hand column below. This is our opinion, based on experience. Since some important factors (absolute permeability, fracturing, etc.) cannot be evaluated by means of logs, our opinions cannot always be correct.

If you desire to check our calculations, we show the analysis methods below.

We hope you will find this information helpful in the completion of your well. If we can be of further service to you, please let us know.

SCHLUMBERGER WELL SURVEYING CORP
R. R. Pagop

| DEPTH | $R_m$ | $R_{mc}$ | $R_{mf}$ | $R_z$ | $R_{f/ \Sigma}$ | SSP | $R_t$ | POROSITY | WATER SAT | POSS PR. |
|---|---|---|---|---|---|---|---|---|---|---|
| 8613-46 | 1.0 | 1.13 | .66 | 8.5 | 5.2 | – | 150 | 12% | 18.5% | OIL OR GAS |

Fig. 59 Log example 11.

### *Log Example 11 Answers*

*For the interval 8618-8647 ft: Porosity = 10-13%; $S_w$ = 17-18%. Net pay footage is about 17-18 ft. The SP is not developed because of the high shale content of the formation. This well made over 400 barrels of oil per day at the start.*

### Log Example 12

**Background:** Muddy ''J'' sand - Denver basin, Colorado.

**Logs:** E-log and MicroLog.

**Given:** $R_m$ = 1.5 ohm-m at 145° F at 6700 ft.

$R_{mf}$ = 1.2 ohm-m at 145° F at 6700 ft.

$R_w$ = 0.1 ohm-m at 145° F at 6700 ft.

Moderate invasion.

**Instructions:** Interval to interpret: 6654-6668 ft.

Using the Rocky Mountain method, compute $S_w$ and porosity.

Check the answer using another interpretation method.

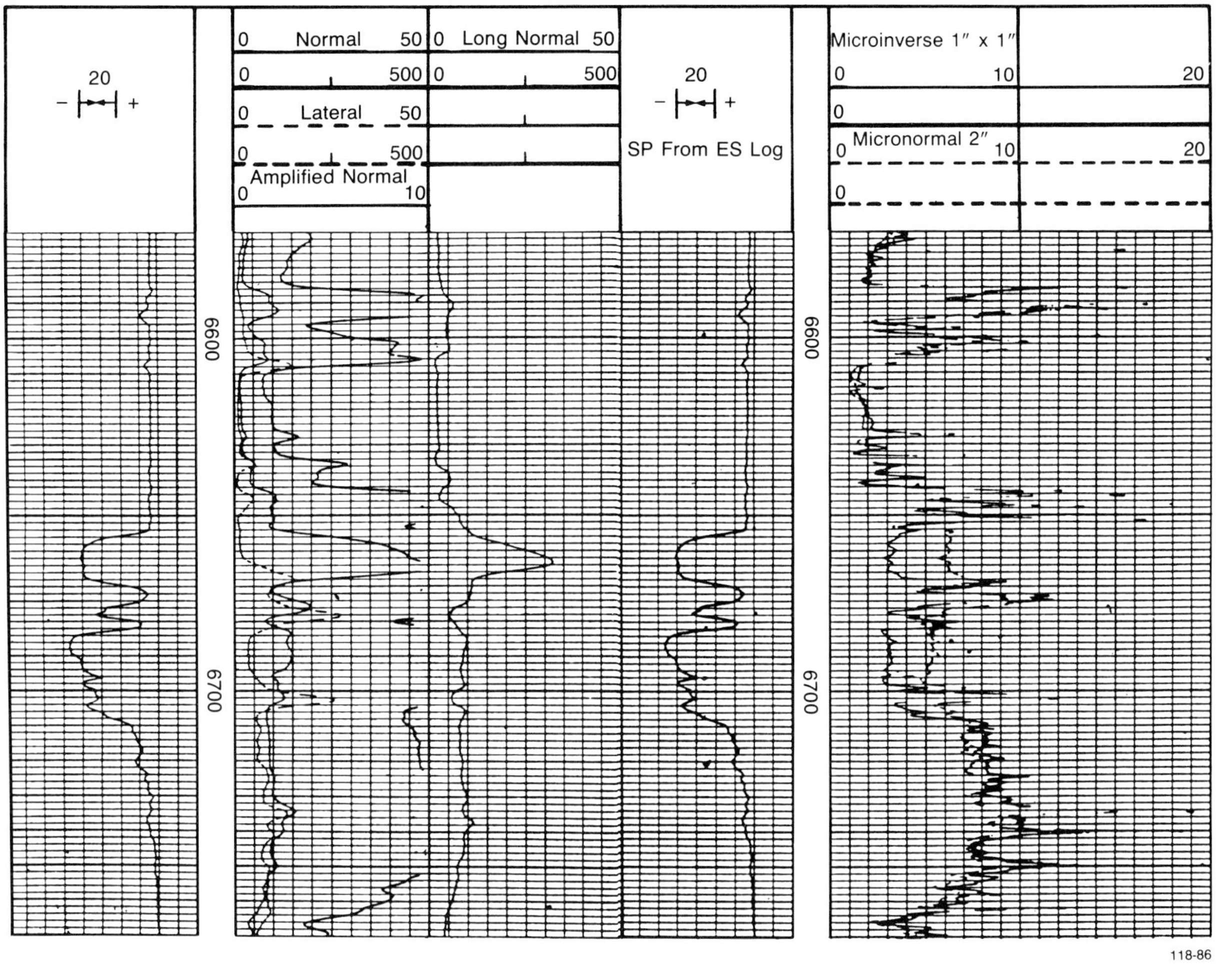


| | | | |
|---|---|---|---|
| Mud Nature | GEL | | |
| " Density | 10.4 | | |
| " Viscosity | 170 | | |
| " Resistivity | 3.6 @ 62°F | @ °F | @ °F |
| " Resistivity BHT | 1.5 @ 145°F | @ °F | @ °F |
| " pH | 8 @ °F | @ °F | @ °F |
| " Water Loss | 7.2 CC/30 Min | CC/30 Min | CC/30 Min |
| Maximum Temp. °F | 145° | | |
| Bit Size | 8¾" TO 4444'; | 7⅞" TO T.D. | |
| Spacings—AM | 16" | | |
| AM' | 64" | | |
| AO | 18'8" | | |

Fig. 60 Log example 12. Courtesy of Douglas W. Hilchie *Old Electrical Log Interpretation* (1979).

### *Log Example 12 Answers*

*Data from the Rocky Mountain method:* $R_{16''} = 50$ *ohm-m;* $R_{64''} = 32$ *ohm-m;* $R_{LAT}$ = *blinded, no value;* $R_m = 1.5$ *ohm-m;* $R_{mf} = 1.2$ *ohm-m; SP = 71 mV; Hole size = 7⅞ in.;* $R_{16''}$ *(corrected) = 45 ohm-m (Chart B-2);* $R_{2''} = 6$ *ohm-m;* $R_i = 45$ *ohm-m;* $R_t = 32$ *ohm-m;* $R_i/R_t = 1.4$*;* $R_t/R_w = 320$*. Plot* $R_i/R_t$ *and* $R_{mf}/R_w$ *on Chart D-8 for z = 0.075 for moderate invasion:* $S_w = 23\%$*. Plot* $R_t/R_w$ *and* $S_w$ *on Chart D-10: Porosity = 21%. Another method: Porosity from the MicroLog using the empirical Chart C-2: 20%;* $S_w$ *from the Archie equation:* $S_w = 28\%$*. Either method indicates a hydrocarbon zone. This zone made a good oil well.*

### Log Example 13

**Background:** Upper cretaceous sand - east Texas.

As the old adage goes, "looks are deceiving." Almost everyone can find the obvious pay zone at 6370-6380 ft, but few recognize the massive sand at 6480-6500 ft as productive. It was passed up for many years as a wet zone until the cuttings were more closely examined for shows. This well produced 53,000 barrels of oil from this interval. Although this sand has very good porosity, it is extremely fine grained with low permeability.

**Logs:** 1954 E-log and MicroLog.

In this example you will find a "mud log" at the bottom of the MicroLog. This is a recording of the resistivity of the mud with the sonde in the collapsed position while going in the hole. Whenever a MicroLog is presented with a mud log, use it for $R_m$.

$R_m$ = 0.380 ohm-m at 6100 ft (174° F).

$R_w$ = 0.019 ohm-m at 6100 ft.

**Instructions:** Interpret the intervals 6370-6380 and 6480-6500 ft using any interpretation method and tabulate the results.

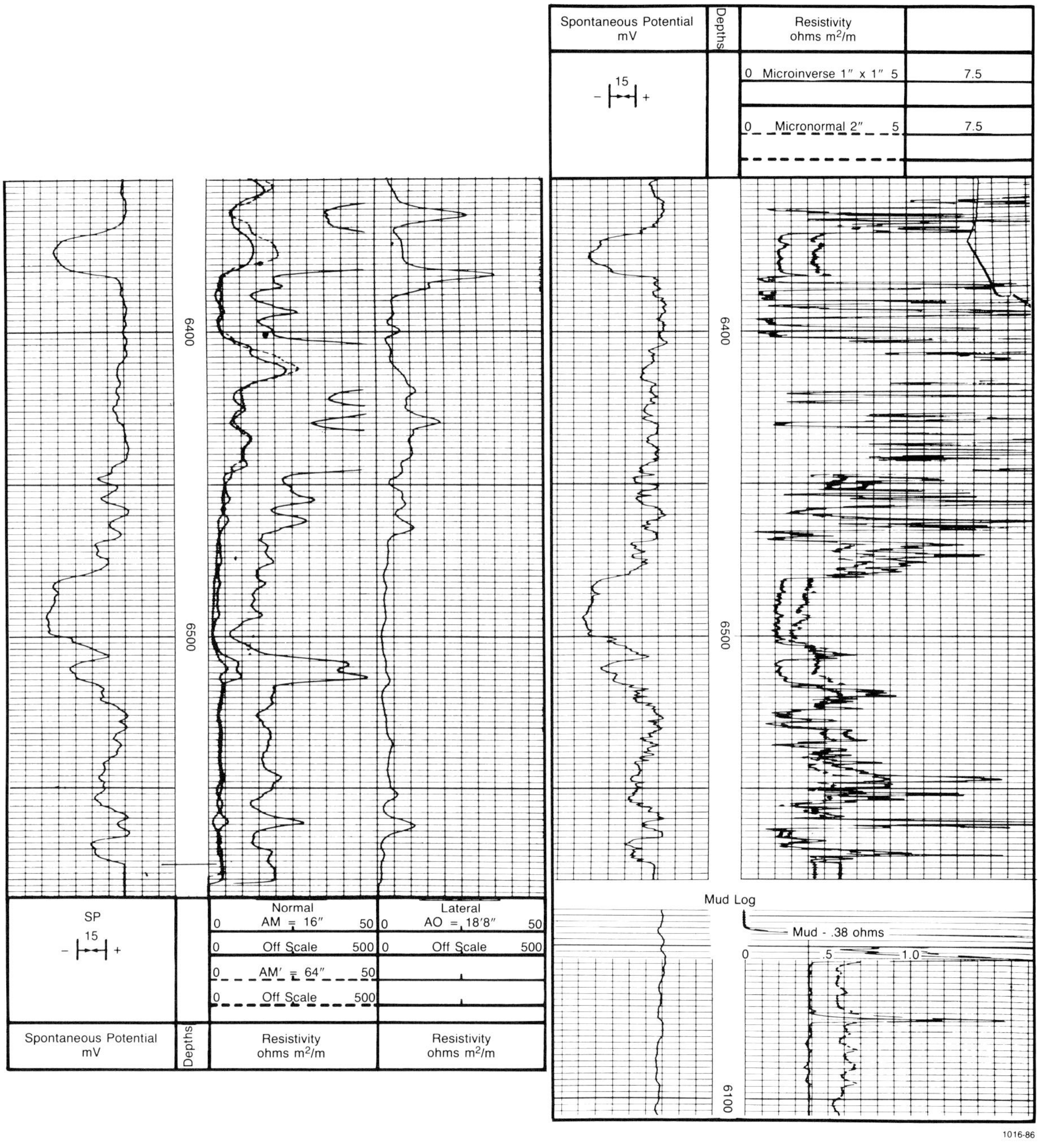


Fig. 61 Log example 13.

### *Log Example 13 Answers*

*For zone 6370-6380 ft: $R_i$ = 14 ohm-m; $R_t$ = 22 ohm-m; $R_{2''}$ = 2.3 ohm-m; $R_{mf}$ = 0.25 ohm-m. Chart C-2 gives porosity = 18% and F = 31 (m = 2); $R_{xo}$ = F $R_{mf}$ = 31 x 0.25 = 7.75 ohm-m; $R_{xo}/R_t$ = 0.35. Chart D-4 gives $S_w$ = 11%. Archie equation: $S_w = \sqrt{F R_w / R_t} = \sqrt{31 \times 0.019/22}$ = 16%. This zone should make a good well. It has not been tested. For zone 6480-6500 ft: $R_{2''}$ = 1.5-2 ohm-m; $R_t$ = 2-2.5 ohm-m (from lateral). Chart C-2 gives porosity = 20-25% and F = 16-25 (m = 2); $R_{xo}$ = F $R_{mf}$ = 25 x 0.25 = 6.25 ohm-m; $R_{xo}/R_t$ = 2.5. Chart D-4 gives $S_w$ = 35%. Archie equation: $S_w \sqrt{F R_w / R_t} = \sqrt{25 \times 0.019/2.5}$ = 44%. This well has produced 53,000 barrels of oil from this interval.*

### Log Example 14

**Background:** Mississippi chat - Kansas.

**Logs:** 1957 Laterolog-7 and Microlaterolog.

**Given:** Diameter of invasion = 20 in.

$R_m$ = 0.057 ohm-m at 4600 ft.

$R_{mf}$ = 0.045 ohm-m at 4600 ft.

$R_w$ = 0.035 ohm-m at 4600 ft.

**Instructions:** Calculate porosity and $S_w$ over the interval 4615-4625 ft.

Does $S_w$ increase below 4625 ft?

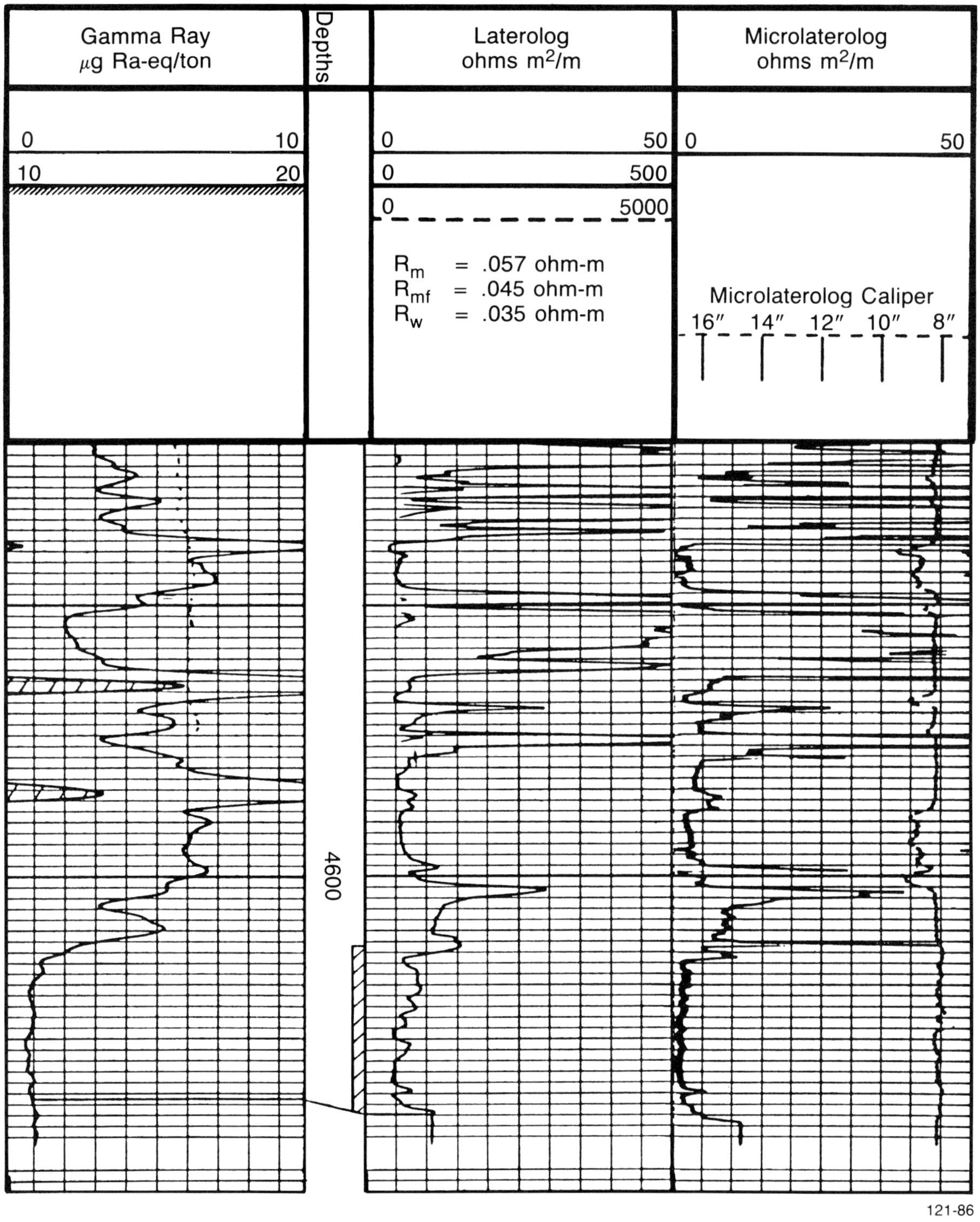


Fig. 62 Log example 14. Courtesy or Douglas W. Hilchie *Old Electrical Log Interpretation* (1979).

### *Log Example 14 Answers*

*Porosity = 16-17%; $S_w$ = 35-37%. Below 4625 ft, $S_w$ shows a slight increase to around 40%. AOF for this well was 9.5 million cubic feet of gas per day over the perforated interval shown on the log.*

### Log Example 15

**Background:** Pennsylvanian limestones - north central Texas.

**Logs:** Openhole gamma ray-neutron logs.

**Given:** GNT-F tool.

19½-in. spacing between source and detector.

PuBe radiation source.

**Instructions:** Using Chart C-18, determine a porosity index for Zones A and B.

What neutron value (in API units) should be used for the minimum porosity control point?

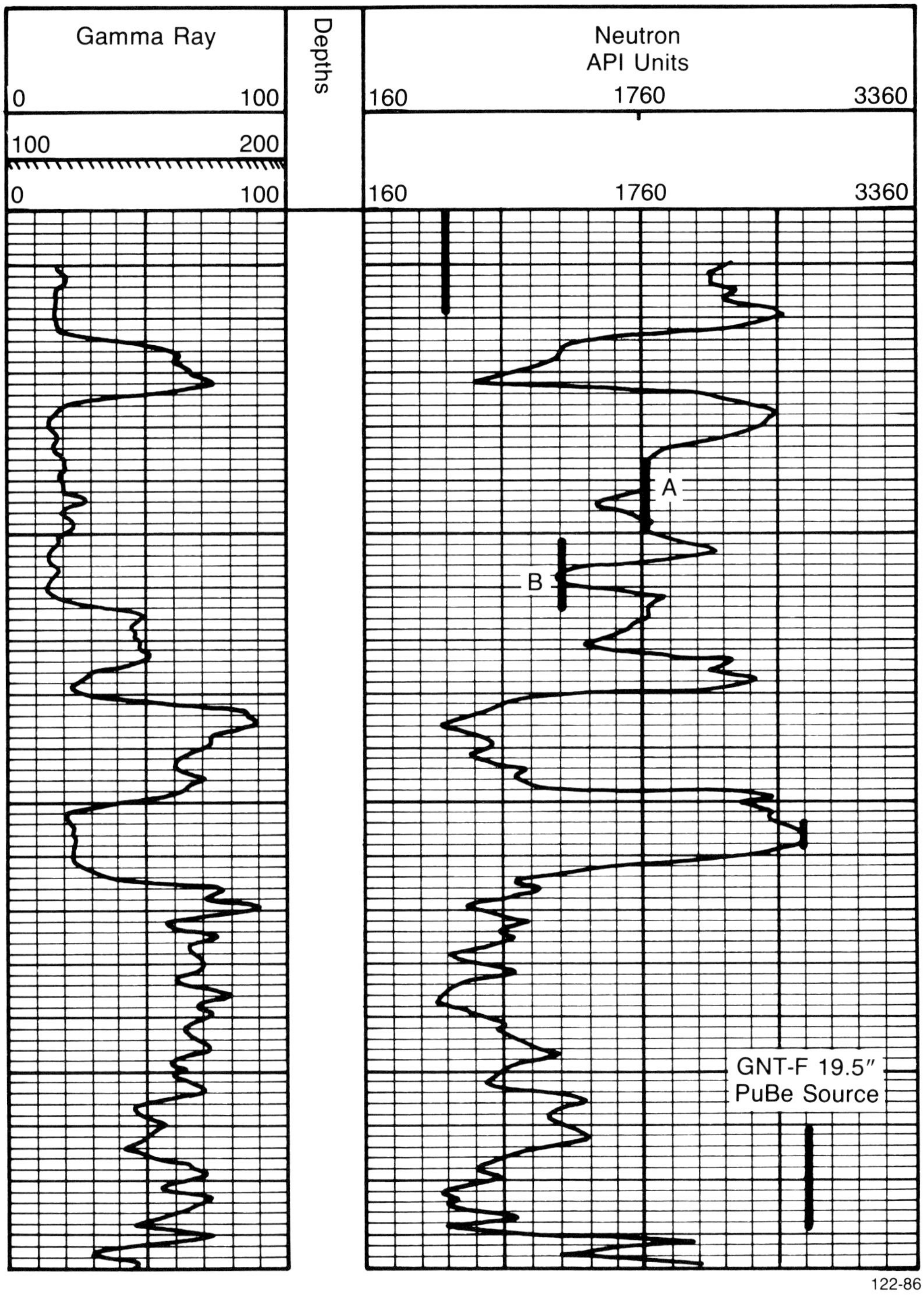


Fig. 63 Log example 15.

### *Log Example 15 Answers*

*The pivot point for this tool is 500 API units and 47% porosity (given on Chart C-18). If the minimum control point of 1% porosity (a tight lime) is selected, the porosity index for Zone A is around 5% and around 12% for Zone B. The minimum porosity control point is approximately 2720 API units (calculated from the log scale). If the minimum control point selected is 2%, the computed porosities would increase nearly 3% in each case. Note the importance of selecting the control points. Local knowledge helps in this decision. A similar interpretation can be made using the logarithmic scaler. Place the shale control point (40%) at the shale baseline and place the minimum control point (1 or 2%) on a tight lime peak. The log is then scaled directly in porosity.*

### Log Example 16

**Background:** Devonian - Alberta, Canada. Top of LeDuc is 7290 ft.

**Logs:** 1959 Laterolog-7, MicroLog, and gamma ray-neutron log.

**Given:** $R_w$ = 0.04 ohm-m at 7300 ft.

$R_m$ = 0.80 ohm-m at 7300 ft (154° F).

The driller reported a show at 7320-7330 ft.

**Instructions:** Should this zone be tested?

Calculate porosity and $S_w$.

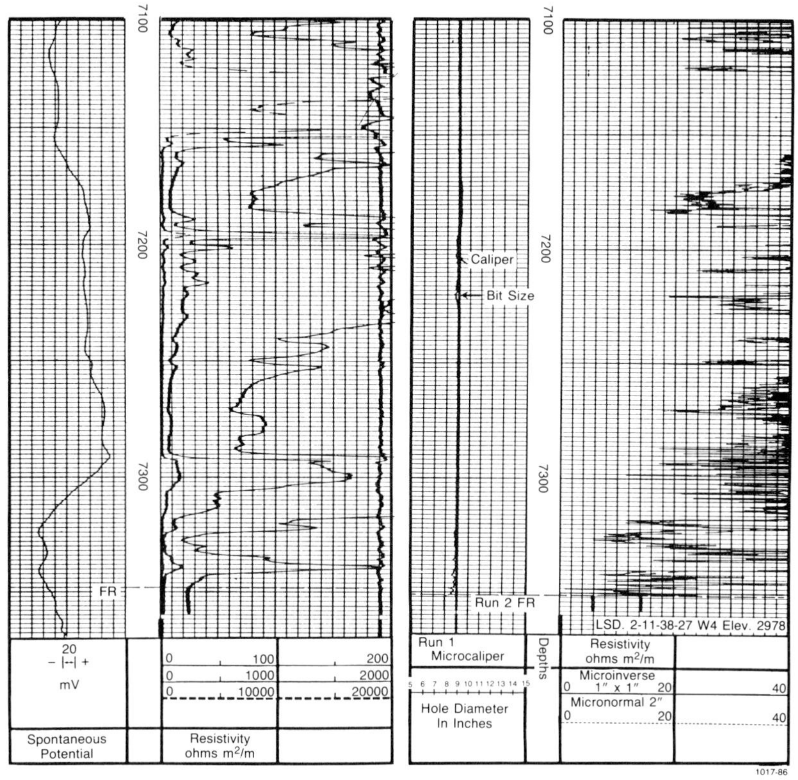


Fig. 64 Log example 16A.

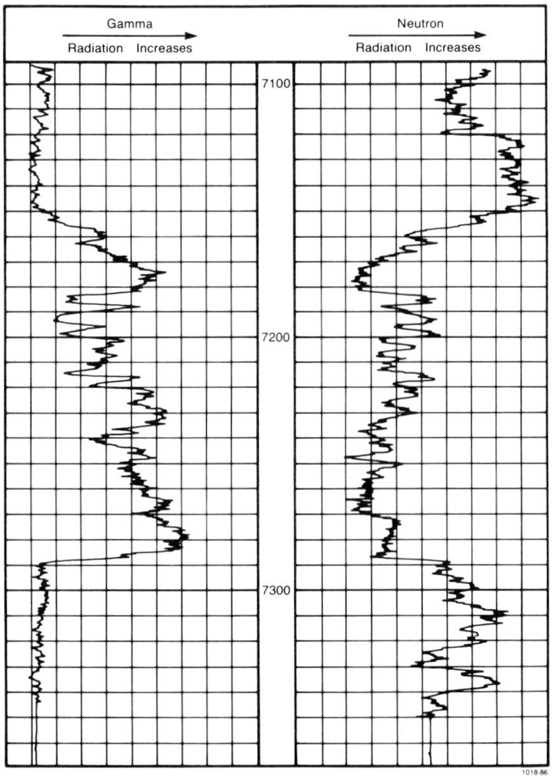


Fig. 65 Log example 16B.

### *Log Example 16 Answers*

*The MicroLog $R_{2''}$ = 12 ohm-m. Chart C-2 gives porosity = 7%. Using the neutron scaler method and selecting nine division deflections for the minimum porosity (1%) and 2.5 division deflections for the shale porosity (30%), the zone of interest has approximately 8% porosity. F(m = 2) = 156 and $R_{LL}$ = 20-30 ohm-m.*

*$S_w = \sqrt{156 \times 0.04/20} = 0.56$ to $\sqrt{156 \times 0.04/30} = 0.45$. This is not worth a test. The Laterolog is probably reading too high since fresh mud filtrate is present; therefore, the $S_w$ will be higher than calculated. The zone at 7290-7305, although tight (less than 5% porosity), could be a fractured zone according to the MicroLog. In Paleozoic carbonates, porosities greater than 2% could be productive when fractured.*

### Log Example 17

**Background:** Lower cretaceous - Alberta, Canada. Top of the Viking sand is 6110 ft.

**Logs:** 1959 induction log and MicroLog.

**Given:** $R_w$ = 0.14 ohm-m at 6100 ft.

$R_m$ = 2.10 ohm-m at 6100 ft (131° F).

**Instructions:** Should this zone be tested?

Calculate porosity and $S_w$.

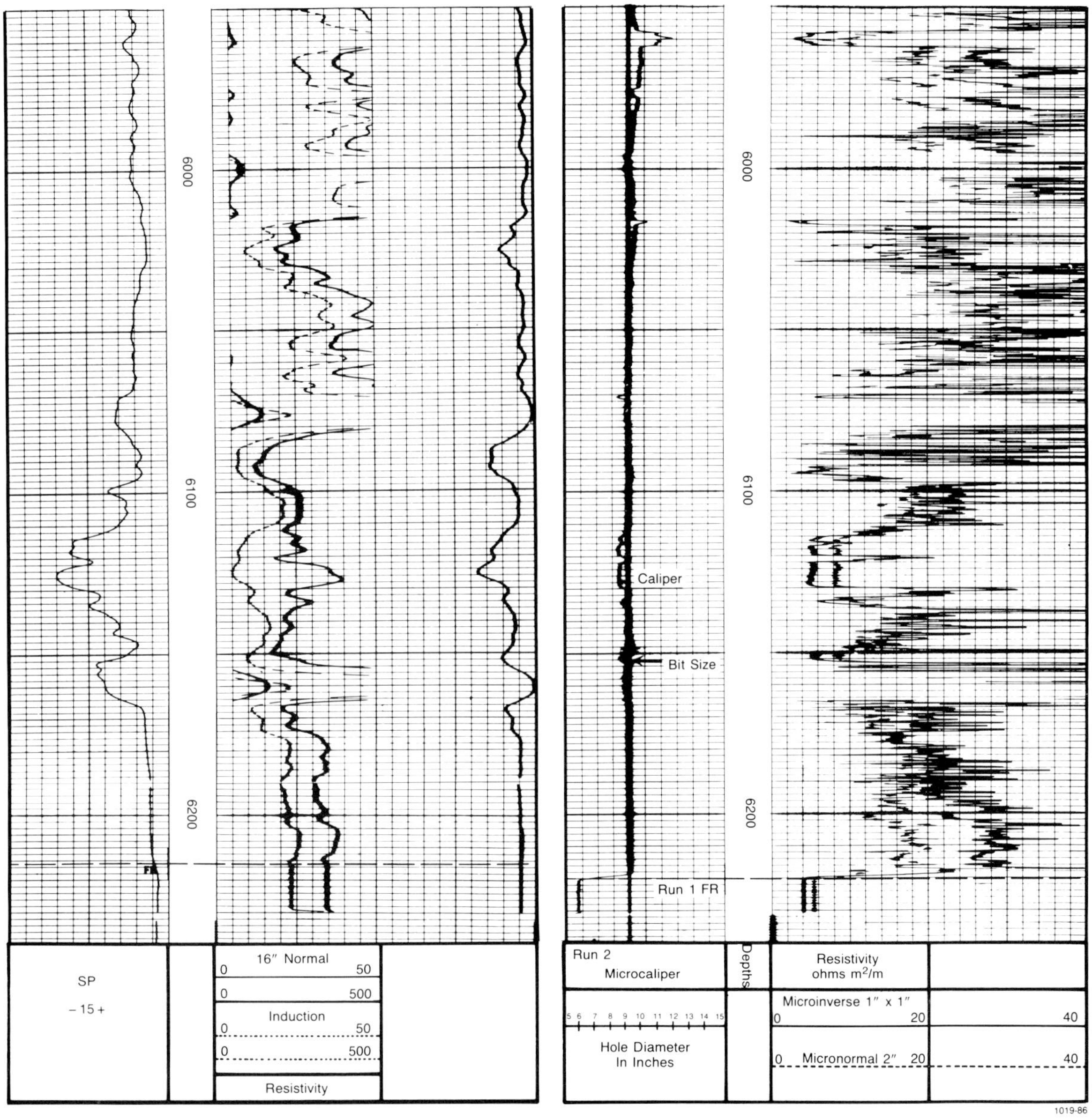


Fig. 66 Log example 17.

### *Log Example 17 Answers*

*For Zone 6114-6420: $R_{mf}$ = 1.8 ohm-m at 131° F; SSP = 86; $R_{2''}$ = 8.3 ohm-m; $R_{1''x1''}$ = 5.2 ohm-m; $R_{mc}$ = 2.1 ohm-m and $R_{mf}/R_w$ = 12.9. Chart C-4 gives $R_{xo}$ = 42 ohm-m. Chart C-2 gives porosity = 19.5% and F = 20. $FR_w$ = 20 x 0.14 = 2.8 and $S_w = \sqrt{2.8/7}$ = 0.63 (slight show), but the zone is shaly; therefore, try the D-4 method. $R_{xo}/R_t$ = 42/7 = 6 and PSP = 67 plots on the 100% line. This could mean that the oil is in the transition zone and not movable. For Zone 6122-6130: MicroLog porosity is about the same (19.5%); $R_t$ = 5 ohm-m; PSP = 83 mV; $R_{xo}/R_t$ = 42/5 = 8.4. Chart D-4 gives very high $S_w$. Neither zone is worth testing.*

# 11 Formation Tester Interpretation

## GENERAL INFORMATION

The more than 50,000 operations run in the 1950's and early 1960's established the formation tester (FT) as an important evaluation tool, and many well log files still contain records from these old wireline tests. A number of papers have been written on applications of formation tester data, and interpretation methods have been developed for various formation and fluid conditions. The purpose of this chapter is to review methods of interpretation to predict fluid production. The charts used in these applications are contained in Appendix B.

Although these interpretation techniques were developed for the old FT tools, they are still valid for interpreting fluid recoveries from modern wireline test tools.

Fig. 67 shows a typical pressure recording obtained with the old FT or formation interval tester tools.

A recovery and interpretation data sheet is shown in Fig. 68. Data sheets of this type were usually included as part of the log.

Interpretation of formation tester data is based on the supposition that the recovery, except for the filtrate, is representative of formation fluids that would be produced from the test zone. This same assumption is the basis for other types of nonproduction tests and is usually valid. However, in some cases (tests made near a water/oil contact, thin lense of restricted lateral extent, tests in which the reservoir hydrodynamic conditions are upset, etc.) the interpretations may not be substantiated by production figures. For this reason, use all available information from well logs, geophysical and geological data, and previous experience when interpreting formation tester results.

## FORMATION FLUID DETERMINATIONS

Ideally, a formation tester recovers a sample chamber full of representative formation fluids. The sample gives a direct measurement of relative quantities of gas, oil, and formation water to be expected from completion in the test interval. No special interpretation methods are required for such a recovery.

However, except in through-casing tests, the recovered sample is usually contaminated with mud filtrate. The difficulty of interpretation varies with the degree of filtrate contamination. Furthermore, in low-permeability formations the sample chamber is often intentionally sealed before it is full — and only small quantities of formation fluids are recovered. As long as mud filtrate represents only a small percentage of the recovered fluids and large recoveries are obtained, interpretation remains simple. But, as the fraction of filtrate increases and the volume of formation fluids becomes smaller, interpretation is less straightforward. Special interpretation techniques are required to extrapolate small formation-fluid recoveries to production predictions, and the interpretations are not as accurate or reliable as those based on large recoveries of formation fluids.

### Large Recoveries of Formation Fluids

Chart FT-1 offers a simple and rapid evaluation of tests in which relatively large volumes of formation fluids are recovered. To use the chart, the gas recovery (in cubic feet) is plotted versus the oil recovery (in cubic centimeter) and the resulting plot is compared with a curve corresponding to the recorded shut-in pressure. A plot above the appropriate shut-in pressure curve is a reliable indication that the tested interval will produce gas and/or oil on initial production. However, this empirical chart tends to be pessimistic under certain conditions and a plot below the shut-in pressure curve is not a positive condemnation of the interval.

The chart was constructed on the basis of tests using a 2¾-gal chamber in relatively high-porosity formations. Invasion in such formations is usually shallow and permeability is relatively high; full-chamber recovery with an appreciable fraction of formation fluids is the rule

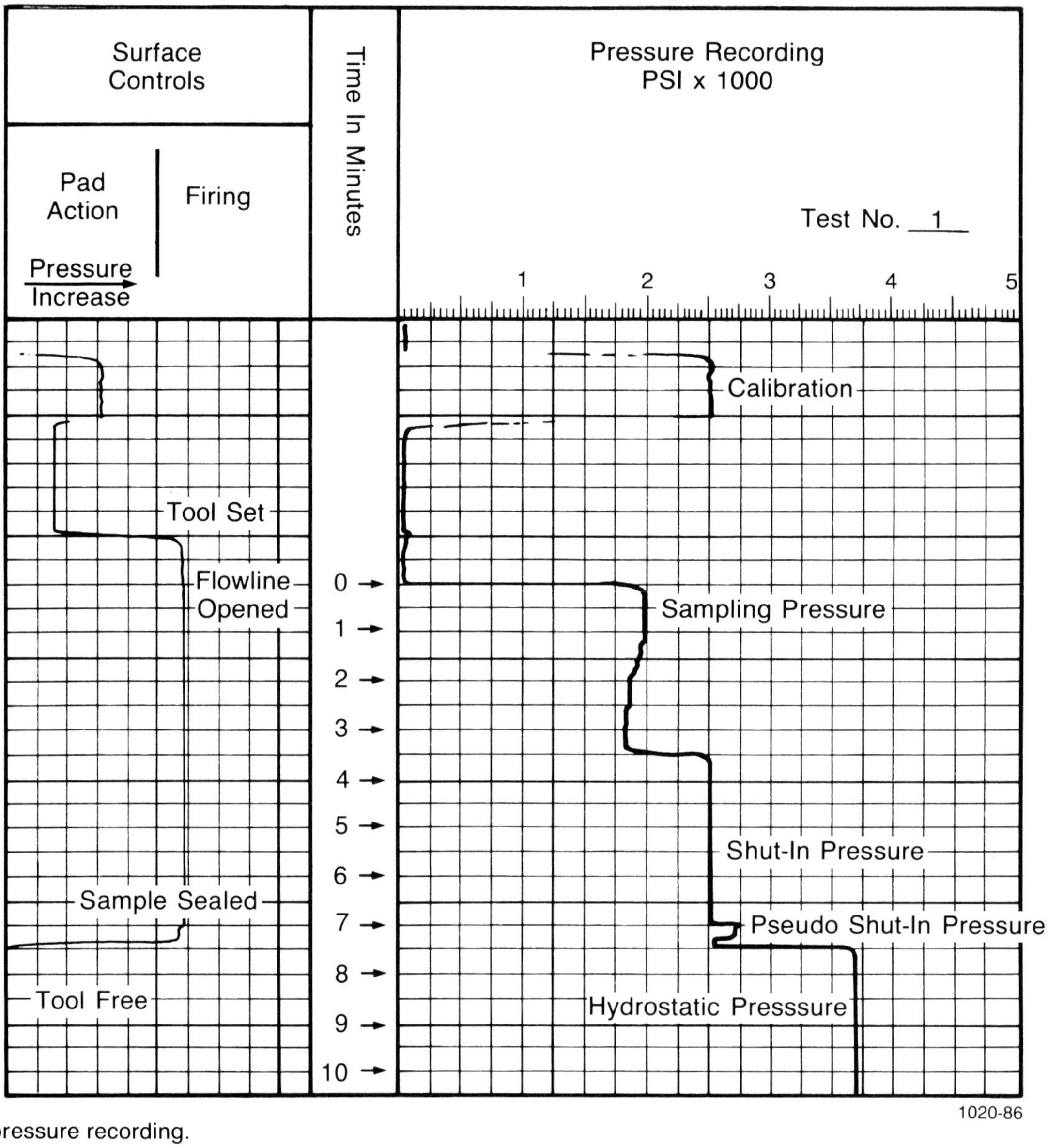


Fig. 67 FT pressure recording.

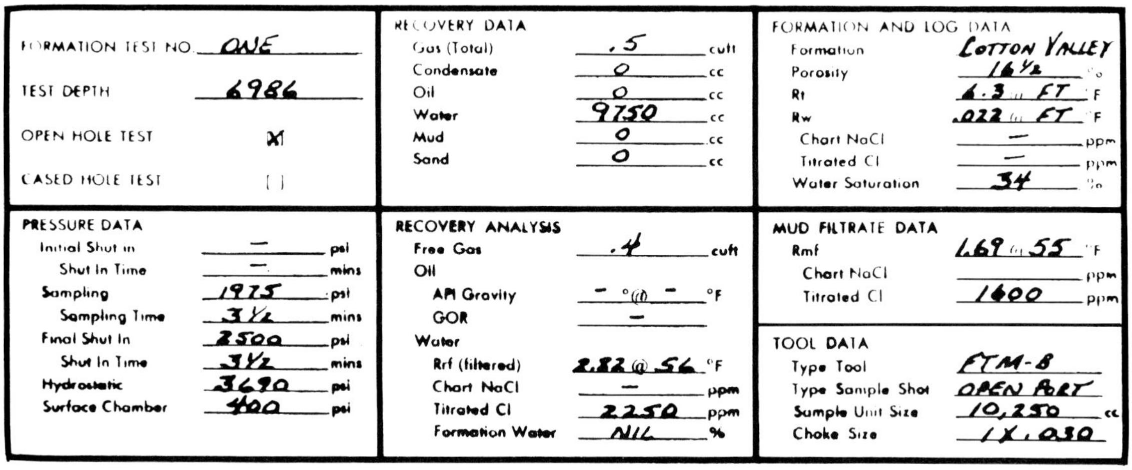

**FORMATION TEST NO.** ONE
**TEST DEPTH** 6986
**OPEN HOLE TEST** [X]
**CASED HOLE TEST** [ ]

**RECOVERY DATA**

| | | |
|---|---|---|
| Gas (Total) | .5 | cuft |
| Condensate | 0 | cc |
| Oil | 0 | cc |
| Water | 9750 | cc |
| Mud | 0 | cc |
| Sand | 0 | cc |

**FORMATION AND LOG DATA**

| | | |
|---|---|---|
| Formation | COTTON VALLEY | |
| Porosity | 16½ | % |
| Rt | 6.3 @ FT | °F |
| Rw | .022 @ FT | °F |
| Chart NaCl | – | ppm |
| Titrated Cl | – | ppm |
| Water Saturation | 34 | % |

**PRESSURE DATA**

| | | |
|---|---|---|
| Initial Shut in | – | psi |
| Shut In Time | – | mins |
| Sampling | 1975 | psi |
| Sampling Time | 3½ | mins |
| Final Shut In | 2500 | psi |
| Shut In Time | 3½ | mins |
| Hydrostatic | 3690 | psi |
| Surface Chamber | 400 | psi |

**RECOVERY ANALYSIS**

| | | |
|---|---|---|
| Free Gas | .4 | cuft |
| Oil | | |
| API Gravity | – °@ – | °F |
| GOR | – | |
| Water | | |
| Rrf (filtered) | 2.82 @ 56 | °F |
| Chart NaCl | – | ppm |
| Titrated Cl | 2250 | ppm |
| Formation Water | NIL | % |

**MUD FILTRATE DATA**

| | | |
|---|---|---|
| Rmf | 1.69 @ 55 | °F |
| Chart NaCl | | ppm |
| Titrated Cl | 1600 | ppm |

**TOOL DATA**

| | | |
|---|---|---|
| Type Tool | FTM-8 | |
| Type Sample Shot | OPEN PORT | |
| Sample Unit Size | 10,250 | cc |
| Choke Size | 1X .030 | |

Fig. 68 Typical FT data sheet.

rather than the exception. However, with deep invasion, such as often occurs in low-porosity formations, the chart becomes pessimistic. With filtrate a high fraction of the total recovery, the volume of formation fluids is low, and a potentially productive oil or gas sand may plot below the appropriate shut-in pressure curve. Thus, additional studies are required before condemning a zone that is not indicated good by Chart FT-1.

A word of caution in using Chart FT-1: the chart is based on full-chamber recovery with a 2¾-gal tester. If a different size chamber is used, or if only partial recovery is obtained, the quantities of fluid must be adjusted until equivalent to 2¾-gal recoveries. For example, if a 5½-gal chamber was used, and if it was full, divide the oil and gas recoveries by two before entering the chart.

Table 5 lists the sample chamber sizes and volumes. The 2¾-gal chamber was the standard unit and the majority of tests were run with this size. The sample unit size used for each test is listed on the log insert or data sheet.

**Table 5**
Sample Chamber Volumes

| | |
|---|---|
| ½-Gal Segregator Chamber | 2,225 $cm^3$ |
| 1-Gal Sample Chamber | 3,690 $cm^3$ |
| 2¾-Gal Sample Chamber | 10,250 $cm^3$ |
| 5½-Gal Sample Chamber | 20,500 $cm^3$ |
| 6-Gal Sample Chamber | 22,165 $cm^3$ |
| 12-Gal Sample Chamber | 44,815 $cm^3$ |
| 18-Gal Sample Chamber | 67,450 $cm^3$ |
| 24-Gal Sample Chamber | 90,080 $cm^3$ |

GOR Determination

The diagonal lines on Chart FT-1 provide an estimate of the production gas-oil ratio expressed in cubic feet of gas per barrel of oil. These lines are a graphical solution of the following expression:

$$\text{GOR} = \frac{\text{cu ft gas recovered}}{\text{cm}^3 \text{ oil recovered}} \times 159{,}000 \text{ cm}^3/\text{bbl} \quad (11\text{-}1)$$

Water-Cut Determinations

Although the amounts and natures of the recovered waters were not used on Chart FT-1, these data should be considered. With relatively large recoveries of formation water, production from the interval will be water cut. The production water cut will be approximately equal to the water cut of formation liquids in the tester recovery and is defined by the following equation:

$$\text{Water Cut} = \frac{\text{cm}^3 \text{ formation water}}{\text{cm}^3 \text{ formation water} + \text{cm}^3 \text{ oil}} \quad (11\text{-}2)$$

Note that the water-cut determination is based upon the amount of formation water. Because sizable quantities of mud filtrate are usually recovered, the total volume of recovered water can seldom be entered in Eq. 11-2. It is necessary to first determine the relative fractions of mud filtrate and formation water and then to solve for the volume of formation water. This determination of the fractions of filtrate and formation water can be accomplished by either titration or resistivity measurements. For the latter approach, Chart FT-2 can be used to determine the fraction of formation water in the tester recovery. The chart is appropriate for fresh muds where $R_{mf}$ is at least $2R_w$; accuracy is improved at higher $R_{mf}/R_w$ values. In using the chart it is important that all fluid resistivities be adjusted to a common temperature. Chart FT-3 is used to adjust the resistivities of the recovered water, the known $R_w$, and mud filtrate to a common temperature datum. After the resistivities are adjusted, a plot of $R_{mf}/R_{rf}$ versus $R_{mf}/R_w$ defines, with respect to the slanted lines, the percentage of the recovered water that is formation water. Multiplying the volume of water recovered by this percentage gives the volume of formation water to use in Eq. 11-2.

It has been observed that when less than 600 $cm^3$ of formation water are recovered along with large quantities of oil and/or gas, production is generally water free. For this reason water-cut computations of less than 10% are usually ignored.

The following example is typical of tests in which hydrocarbon production is predicted on the basis of Chart FT-1:

Frio formation, Newton County, Texas

| | | |
|---|---|---|
| Test depth: | 7560 ft. | |
| Tool data: | 2¾-gal sample chamber. | |
| Pressures: | Sampling | 2,600 psi. |
| | Shut-in | 2,800 psi. |
| Recovery: | Gas | 23 cu ft. |
| | Oil | 7,250 $cm^3$. |
| | Water | 350 $cm^3$. |
| | $R_{rf}$ | 0.88 at 78° F. |
| | Sand | 200 $cm^3$. |
| Log Data: | $R_{mf}$ | 1.90 at 78° F. |
| | $R_w$ | 0.32 at 78° F. |

The plot on Chart FT-1 of gas versus oil indicates the interval will produce oil, with a GOR of 550:1. With the resistivity of the recovered water less than that for mud filtrate, there is evidence some formation water was recovered. Chart FT-2 indicates 32% of the recovered water is formation water. Thus, of the 350 $cm^3$ of recovered water, 112 $cm^3$ are formation water and 238 $cm^3$ are mud filtrate. However, with the large hydrocarbon recovery, the 112 $cm^3$ of formation water is not considered significant and the zone was expected to produce water-free oil.

Production results: This zone was perforated from 7552 to 7562 ft, or from 8 ft above to 2 ft below the tested interval. The zone produced 120.9 barrels of oil per day on a ⅛-in. choke, with a GOR of 710:1, and with no water.

In these tests the oil and gas recoveries were large and the plotted point fell well above all the shut-in pressure curves. This result is typical of tests in shallow-invaded, high-porosity, high-permeability oil or gas sands. Indeed, a glance at Chart FT-1 shows that the tested interval should produce oil and/or gas if more than 10 cu ft of gas or more than 1,100 $cm^3$ of oil are recovered (when using a 2¾-gal chamber).

Experience shows that tests in high-porosity, high-permeability sands rarely lead to plots within the family of shut-in pressure curves. With the relatively large volumes of formation fluids recovered from such sands, the plotted points usually fall well above the pressure curves in oil- or gas-productive intervals or well below in water-productive sands. Therefore, if a large fraction of the recovered water is formation water, and if small hydrocarbon recoveries lead to a plot in the "water zone" on the chart, the tested interval may be confidently condemned. Note, however, that the zone may be condemned with confidence only when large fractions of formation water are recovered.

A test leading to a plot in the water zone of Chart FT-1 should not be considered conclusive when the recovered water is predominantly mud filtrate.

**Small Recoveries of Formation Fluids**

When the fluid recovered in a test is predominantly mud filtrate, even a very small show of oil or gas may be significant. With deep invasion, such as most commonly found in low-porosity formations, most of the formation fluids are flushed from the zone close to the borehole. Therefore, a formation test in a deeply invaded sand may recover a sample chamber full of mud filtrate.

However, in oil or gas sands flushing is generally never complete — some hydrocarbons remain. When the zone is tested with the formation tester, the formation is opened to approximately atmospheric pressure. The resulting pressure differential causes some of the remaining hydrocarbons to flow into the tool along with the mud filtrate. Therefore, some indication of hydrocarbons is expected in the recovered fluid if the formation is gas bearing and/or oil bearing. However, because these residual hydrocarbons may represent only a small fraction of the recovered volume of fluid, it is important to examine the recovery very carefully.

Recovery of any show of oil in a chamber full of mud filtrate may be significant. Sometimes the quantity of oil is so small it can be detected only by checking for fluorescence. Such checks should be made on all filtrate recoveries.

Recovery of any free gas also indicates the zone may be productive. However, the recovered gas must be examined carefully to determine what fraction of the recovered gas is truly free gas.

All FT shaped charges produce explosion gas. The amount of explosion gas, or "shot gas," recovered depends on the number and the size of the charges. The ⅝-in. shaped charge produces 0.5 cu ft and the ⅞-in. charge produces 1 cu ft. This shot gas must be subtracted from the total gas recovery.

Gas is, to a limited extent, soluble in water. The amount of gas that can exist in this dissolved state is dependent on temperature, formation shut-in pressure, and formation water salinity. Charts FT-4 and 5 give the maximum amounts of gas that the formation water can contain under a given set of conditions. In interpretations of formation test data it is assumed that only formation water contains dissolved gas; so only the amount of formation water recovered is used to determine the maximum dissolved gas. The dissolved gas must also be substracted from the total gas recovered. The remainder after subtracting the shot gas and the dissolved gas is considered free gas.

In addition to the oil and free gas recovered, the recovered water is significant. It is important to know what fraction of the recovered water is formation water. This fraction can be determined by using Chart FT-2.

- If the fraction of formation water recovered is less than 10% of the total water recovered and if, of course, free gas and/or oil was also recovered, the tested interval will produce oil and/or gas with little or no water cut.
- If the recovered formation water fraction is greater than 10% but less than 20-25%, some water cut on production can be expected.
- If the recovered formation water fraction exceeds 20-25%, the production water cut will be excessive.

The following example is typical of the tests in which only small volumes of formation fluids are recovered.

| | | |
|---|---|---|
| Tool data: | 1-gal sample chamber. | |
| | Explosion gas | 0.5 cu ft. |
| Pressures: | Sampling | nil. |
| | Shut-in | 1,500 psi. |
| | Hydrostatic | 2,250 psi. |
| Recovery: | Gas (total) | 1 cu ft. |
| | Oil | 200 $cm^3$. |
| | Water | 3,350 $cm^3$. |
| | $R_{rf}$ | 0.90 at 100° F. |
| Log Data: | SP | −90 mV at 135° F. |
| | $R_{mf}$ | 0.83 at 100° F. |

In this example the resistivity of the recovered fluid ($R_{rf}$) is greater than the resistivity of the mud filtrate ($R_{mf}$) so no formation water was recovered. Since no formation water was recovered, it is assumed that no dissolved gas was present. However, 0.5 cu ft of explosion gas was produced. Thus, the free gas is

$$\text{Free Gas} = 1 - 0.5 = 0.5 \text{ cu ft.}$$

This test indicates water-free production at a GOR of

$$GOR = \frac{0.5 \times 159,000}{200} = 398/1.$$

### Combining FT and Resistivity Information for Improved Interpretations

This $J_{FT}$ versus $R_t/R_w$ technique was developed to use when the data provided by the formation tester are too meager to permit a conclusive interpretation by other methods. In this technique the tester data are combined with information from the resistivity and SP logs to improve the interpretation.

$J_{FT}$ is determined as follows:

$$J_{FT} = \frac{0.55q}{h\ (p_{ws} - p_s)} \qquad (11\text{-}3)$$

where

$J_{FT}$ is the specific productivity index, in B/D/psi-ft;

$q$ is the flowrate into the tester, in cm$^3$/sec;

$p_{ws}$ is the shut-in pressure, in psi;

$p_s$ is the sampling pressure, in psi;

and

$h$ is the thickness of the interval tested (0.75 ft/shot for the FT tool).

Values of $p_{ws}$ and $p_s$ are obtained from the pressure recording of the test. The value of $q$ can be computed by either Chart FT-6 or FT-7. FT-6 is more convenient for high flowrates, while FT-7 is more applicable to lower rates. With the value of $q$ determined, $J_{FT}$ can be found using Chart FT-8.

With $J_{FT}$ from the tester data and $R_t$ and $R_w$ from the resistivity log and the SP, a point can be plotted on Chart FT-9.

- If this point falls in the upper right portion of the chart, a commercial well is indicated.
- If the point falls in the lower left portion of the chart, a noncommercial well is suggested.
- Intermediate points are interpreted as shown on the chart.
- Low values of $J_{FT}$ (less than 0.0003) indicate limited flow potential.

Chart FT-9 was empirically derived from north Louisiana well data, but experience shows that it works well in most areas of North America. However, the lines separating hydrocarbon production from water may not be exactly the same for other areas.

The following example illustrates the use of the $J_{FT}$ interpretation technique.

| | | |
|---|---|---|
| Tool data: | 1-shot FT. | |
| | 5½-gal sample chamber. | |
| | Choke size | 0.030 in. |
| Pressures: | Sampling | 3,650 psi. |
| | Shut-in | 3,900 psi. |
| Recovery: | Gas | 0. |
| | Oil | Scum. |
| | Water | 18,000 cm$^3$ (filtrate). |
| Log Data: | $R_t$ | 2.3 ohm-m. |
| | $R_w$ | 0.023 ohm-m at formation temperature. |

Entering Chart FT-7 with the tester data gives $q$ = 48 cm$^3$/sec from Chart FT-8; $J_{FT}$ = 0.13; $R_t/R_w$ = 100. Plotting these coordinates on Chart FT-9 gives a point well to the right of the dashed line separating the zone of hydrocarbon recovery from the zone indicated as "hydrocarbons with low water fraction." This indicates a commercial well. On production the well produced 132 barrels of clean oil per day.

## INTERPRETATION EXAMPLES

### FT Example 1

**Background:** Miocene formation, Iberville Parish, Louisiana.

**Given:**

| | | |
|---|---|---|
| Test depth: | 9294 ft. | |
| Tool data: | 1-gal sample chamber. | |
| Pressures: | Sampling | 4,300 psi. |
| | Shut-in | 4,400 psi. |
| Recovery: | Gas | 7.5 cu ft. |
| | Oil | 874 cm$^3$. |
| | Water | 1,420 cm$^3$. |
| | $R_{rf}$ | 1.9 at 100° F. |
| Log Data: | $R_{mf}$ | 1.8 at 100° F. |
| | $R_w$ | 0.08 at 100° F. |

**Instructions:** Use the appropriate FT interpretation charts to predict the production.

What will the well produce?

### *FT Example 1 Answers*

*Because this test was made with a 1-gal tester, the volumes of gas and oil must be multiplied by 2.75 before entering Chart FT-1. The chart indicates the tested interval should produce oil, with a GOR of 1500:1. The recovered water was apparently all mud filtrate because $R_{rf}$ was slightly higher than $R_{mf}$. A second test was made in the same zone, but 4 ft shallower. This test also indicated water-free oil production, but with an increased GOR of 3500:1.* ***Production Data:*** *The sand was perforated from 9296 to 9300 ft and, on a DST open for 5¾ hours, flowed 211 barrels of oil on a ⅛-in. choke. The GOR was 1284:1, and final bottomhole flowing pressure was 4,375 psi. The formation was perforated below the lower test level to reduce gas production.*

**FT Example 2**

**Background:** Upper Morrow sand, Beaver County, Oklahoma.

**Given:**

| | | |
|---|---|---|
| Test depth: | 7240 ft. | |
| Tool data: | 5½-gal sample chamber. | |
| Pressures: | Shut-in | 2,250 psi. |
| Recovery: | Gas | 85 cu ft. |
| | Oil | 200 $cm^3$ (condensate). |
| | Water | 6,100 $cm^3$ (filtrate). |

**Instructions:** Use the appropriate FT interpretation charts to predict the production.

What will the well produce?

***FT Example 2 Answers***

*The resistivity of the recovered water was slightly greater than $R_{mf}$, which indicated an absence of formation water. The volumes of gas and oil are halved (to account for the larger volume of the 5½-gal chamber) before entering Chart FT-1. The chart indicates the interval to be gas productive, with a GOR of 65,000:1.* ***Production Data:*** *Although this well was actually completed in the Lower Morrow, from 7212 to 7320 ft was tested conventionally (DST) and flowed 13.3 million cubic feet of gas per day. Eleven feet of condensate were recovered.*

**FT Example 3**

**Background:** Cotton Valley sand, north Louisiana.

**Given:**

| | | |
|---|---|---|
| Tool data: | 2¾-gal sample chamber. | |
| Pressures: | Sampling | nil. |
| | Shut-in | 1,500 psi. |
| | Hydrostatic | 2,500 psi. |
| Recovery: | Gas | 0.9 cu ft. |
| | Oil | 500 $cm^3$. |
| | Water | 9,600 $cm^3$. |
| | $R_{rf}$ | 0.12 at 75° F. |
| Log Data: | SP | −80 mV at 150° F. |
| | $R_{mf}$ | 0.85 at 75° F. |

**Instructions:** Use the appropriate FT interpretation charts to predict production.

What will the well produce?

What are your recommendations?

***FT Example 3 Answers***

*In this test the sizable oil recovery and the presence of free gas suggest that this would be a producing well. However, the difference between the resistivities of the recovered fluid and the mud filtrate indicates that some formation water was recovered. Entering Chart FT-2 with the SP of −80 mV and the ratio ($R_{mf}/R_{rf}$) of 7 gives 60% as the fraction of formation water recovered. The actual amount of formation water recovered is about 5,800 $cm^3$ (60% of 9,600 $cm^3$) and the expected water cut is 5,800/5,800 + 500 x 100 = 91%. In most cases, a well with this high a water cut would be considered noncommercial. Using Chart FT-1, the plotted point lies near the transition point and confirms a probable high water cut. The zone was not completed.*

**FT Example 4**

**Given:**

| | | |
|---|---|---|
| Tool data: | 1-shot FT. | |
| | 1-gal sample chamber. | |
| | Choke size | 0.020 in. |
| Pressures: | Sampling | 500 psi. |
| | Shut-in | 3,550 psi. |
| | Tool open | 6 min. |
| Recovery: | Gas | Trace. |
| | Oil | Scum. |
| | Water | 3,150 $cm^3$ (filtrate). |
| Log Data: | $R_t$ | 2.3 ohm-m. |
| | $R_w$ | 0.025 ohm-m at formation temperature. |

**Instructions:** Use the appropriate FT interpretation charts to predict the production.

What will the well produce?

Would you set pipe on this zone?

***FT Example 4 Answers***

*With this low a sampling pressure, Chart FT-7 is most convenient for determining q. In this example it gives a value of q = 8.7 $cm^3$/sec. The $J_{FT}$ nomogram gives $J_{FT}$ = 0.002 and $R_t/R_w$ = 92. The point described by these coordinates on Chart FT-9 is in the zone labeled "hydrocarbons with high water fraction." On production this well produced water with only a slight show of oil.*

# A

# Log Interpretation Charts for Old Electric Logs

## CONTENTS

SCHLUMBERGER WELL SURVEYING CORP.

# LOG INTERPRETATION CHARTS

## REFERENCE SHEET

## SYMBOLS

AM — Normal Curve Spacing
AO — Lateral Curve Spacing
BHT — Bottom Hole Temperature in °F
d — Diameter of the Hole
Di — Average Diameter of Invaded Zone
e — Bed Thickness in Feet
$t_{mc}$ — Mud Cake Thickness

$R_m$ — Resistivity of the Mud
$R_{mf}$ — Resistivity of the Mud Filtrate
$R_{mc}$ — Resistivity of the Mud Cake
$R_w$ — Resistivity of the Formation Water
$R_{wa}$ — Apparent Resistivity of the Formation Water in Shaly Sands
$R_z$ — Resistivity of the Mixture of the Electrolytes (It can be defined for a given zone — invaded or noninvaded)

$R_t$ — Resistivity of the Formation — Uncontaminated Zone
$R_o$ — Resistivity of the Formation when 100% Water
$R_i$ — Resistivity of the Invaded Zone
$R_{xo}$ — Resistivity of the Flushed Zone (Close to Bore Hole)
$R_s$ — Resistivity of the Surrounding Beds
$R_{16''}$, $R_{64''}$, $R_{18'8''}$, $R_{1''x1''}$, and $R_{2''}$ — Apparent Resistivity of 16″ Normal, 64″ Normal, 18′8″ Lateral, 1″x1″ Microinverse, and 2″ Micronormal, respectively

F — Formation Resistivity Factor
$F_a$ — Apparent Formation Resistivity Factor
$\phi$ — Effective Porosity in Per Cent
$S_w$ — Water Saturation, Per Cent of Pore Space in Uncontaminated Zone
$S_i$ — Water Saturation, as above, in Invaded Zone
$S_{xo}$ — Water Saturation, as above, in Flushed Zone
ROS — Residual Oil Saturation as Per Cent of Pore Space; $= (1 - S_{xo})$

K — Coefficient in the SP Formula
$E_c$ — SP Electrochemical Component
$E_k$ — SP Electrofiltration Component
SSP — Static Spontaneous Potential — The Maximum Possible for a particular $R_{mf}/R_w$
PSP — Pseudostatic Spontaneous Potential — The SP Found in a Thick Shaly Sand
$\alpha$ — SP Reduction Factor — PSP/SSP
k — Permeability in millidarcies

## FORMULAS

**Archie's Formula:** $S_w^n = F\,R_w/R_t$; $S_w = \sqrt[n]{R_o/R_t}$. (n usually taken as 2)

**Formation Resistivity Factor:** $F = R_o/R_w$; $F_a = R_{xo}/R_{mf}$.

**Humble Formula:** $F = 0.62/\phi^{2.15}$.

**SP Formulas:** $SSP = -K \log_{10} R_{mf}/R_w$; $PSP = -K \log_{10} R_{mf}/R_{wa}$.

**$R_i$ Method:** $S_w = \dfrac{F}{a\,R_i/R_m}$ ; $S_w \simeq \dfrac{R_{xo}}{a\,R_i}$.

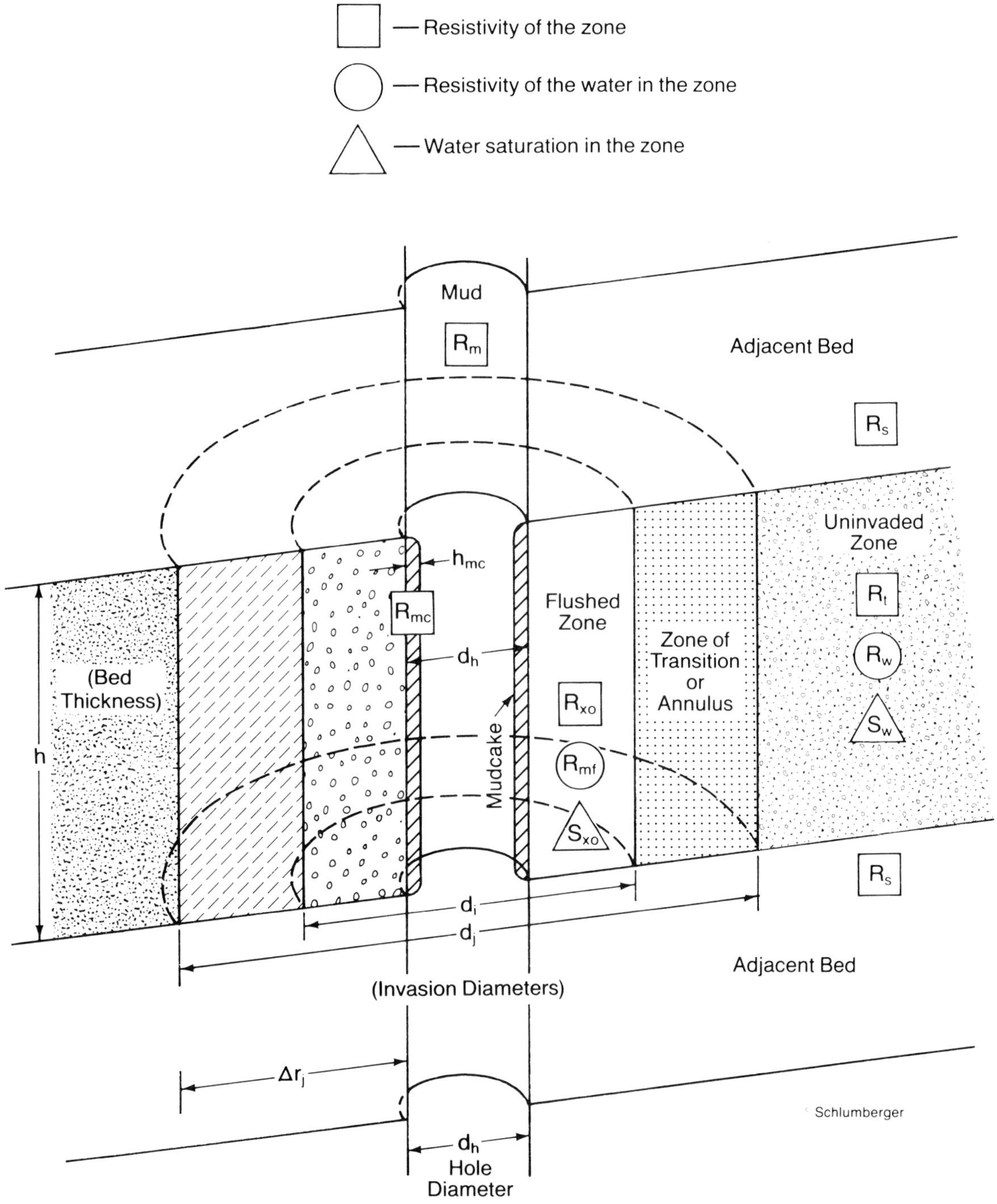
— Resistivity of the zone
— Resistivity of the water in the zone
— Water saturation in the zone
Mud
$R_m$
Adjacent Bed
$R_s$
Uninvaded Zone
$R_t$
$R_w$
$S_w$
$h_{mc}$
$R_{mc}$
Flushed Zone
Zone of Transition or Annulus
$d_h$
(Bed Thickness)
h
$R_{xo}$
Mudcake
$R_{mf}$
$S_{xo}$
$R_s$
$d_i$
$d_j$
Adjacent Bed
(Invasion Diameters)
$\Delta r_j$
Schlumberger
$d_h$
Hole Diameter

SCHLUMBERGER WELL SURVEYING CORP.

# LOG INTERPRETATION CHARTS

## DETERMINATION OF FORMATION TEMPERATURE

### *USE AND DESCRIPTION*

This chart is used:

1. To find the formation temperature at a given depth when it is known at some other depth — usually bottom hole.
2. To estimate temperature at any depth when no temperature has been measured.

The range of normal temperature gradients is shown by diagonal lines. The formation temperatures are scaled according to the mean surface temperature, A. Two values are provided.

### *INFORMATION REQUIRED*

*First Use*

Depth of Formation

Total Depth

Bottom Hole Temperature (BHT)

*Second Use*

Depth of Formation

A — Mean Surface Temperature

G — Geothermal Gradient

### *PROCEDURE*

*First Use*

BHT is known, being 200°F at 11000′. We need to find the temperature at 8000′. Use the upper temperature scale (A = 80). Proceed vertically to intersect the line corresponding to 11000′. Move this point diagonally upward, guided by the gradient line, until it intersects the horizontal line corresponding to 8000′. Read 166°, the temperature corresponding to this point.

*Second Use*

If no bottom hole temperature (BHT) is available, the temperature any place in the hole can be estimated from this chart providing the mean surface temperature, A, and the gradient, G, are known. The relation is of the form: $T^{\circ} = A + (G \times \text{Depth}/100')$. Usually A is about 80° and G is between 0.8 and 1.2 degrees/100′.

EXAMPLE: Formation Depth = 8000′; A = 80°; G = 1.2. Proceed from the depth to the diagonal line of 1.2, go to bottom, and read 176°F on scale starting with 80°.

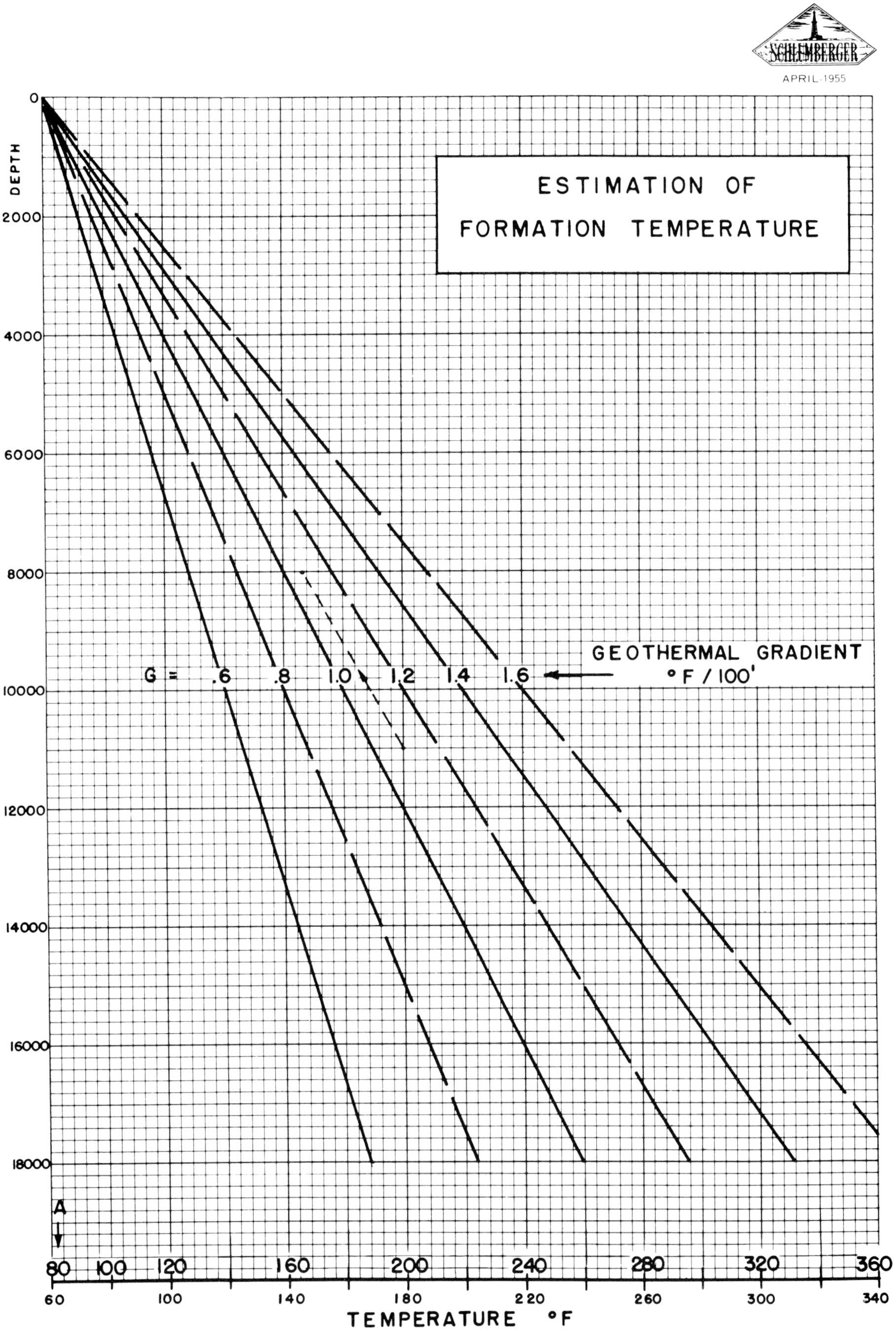
SCHLUMBERGER
APRIL 1955
ESTIMATION OF
FORMATION TEMPERATURE
DEPTH
0
2000
4000
6000
8000
10000
12000
14000
16000
18000
G = .6 .8 1.0 1.2 1.4 1.6
GEOTHERMAL GRADIENT
°F / 100'
A
80 100 120 160 200 240 280 320 360
60 100 140 180 220 260 300 340
TEMPERATURE °F

Chart A-2

SCHLUMBERGER WELL SURVEYING CORP.

# LOG INTERPRETATION CHARTS

## $R_{mc}$ AND $R_{mf}$ DETERMINATION*

*USE AND DESCRIPTION*

This chart is used to find the mud cake and mud filtrate resistivities at formation temperature.

The average values of resistivity of mud cake and filtrate as determined in the laboratory have been plotted versus $R_m$ at formation temperature. $R_{mf}$ is needed to estimate $R_w$ (Charts A-10, D-2). $R_{mf}$ and $R_{mc}$ are both needed to determine F and $\phi$ from MicroLog or MicroLaterolog curves.

*INFORMATION REQUIRED*

$R_m$ at Formation Temperature

Formation Temperature

*PROCEDURE*

From a log heading read $R_m$ at BHT. If several hundred feet above bottom, estimate correct formation temperature, using Chart A-2. Then, $R_m$ at formation temperature = (BHT/Formation Temperature) × $R_m$ at BHT. (Chart A-6 can be used instead.) Enter this on the vertical axis on the left.

Proceed to the right to intersect the dashed curve and read $R_{mf}$ on the horizontal axis. To find $R_{mc}$, go right to intersect the solid curve for the appropriate formation temperature; then read the value on the horizontal axis. For salty muds of resistivity less than 0.1 ohm use the insert chart at right and enter the $R_m$ on the right vertical axis. Find the $R_{mf}$ by going to the dashed line and then down. Find the $R_{mc}$ by going to the solid line and then down.

*EXAMPLE:*

GIVEN: $R_m$ at BHT = 0.4 ohm-m. at 200°F (11000′ depth).

TO FIND: $R_{mf}$ and $R_{mc}$ at 8000′ depth.

SOLUTION: From Chart A-2, temperature at 8000′ = 166°F.

$R_m$ at 8000′ = (200/166) × 0.4 ohm-m. = 0.48 ohm-m. at 166°F.

$R_{mf}$ = 0.30.

$R_{mc}$ = 0.83.

* *IMPORTANT REMARKS:*

$R_{mc}$ and $R_{mf}$ should preferably be measured from mud samples. When these measurements are not available, values can be estimated rather closely by the relations $R_{mc}$ = 1.5 $R_m$ and $R_{mf}$ = 0.75 $R_m$, particularly in muds where NaCl is the major dissolved solid. For other cases, Chart A-4 may be used as a first approximation.

A method of measuring $R_{mc}$ in the hole is described on Charts C-4 and C-6.

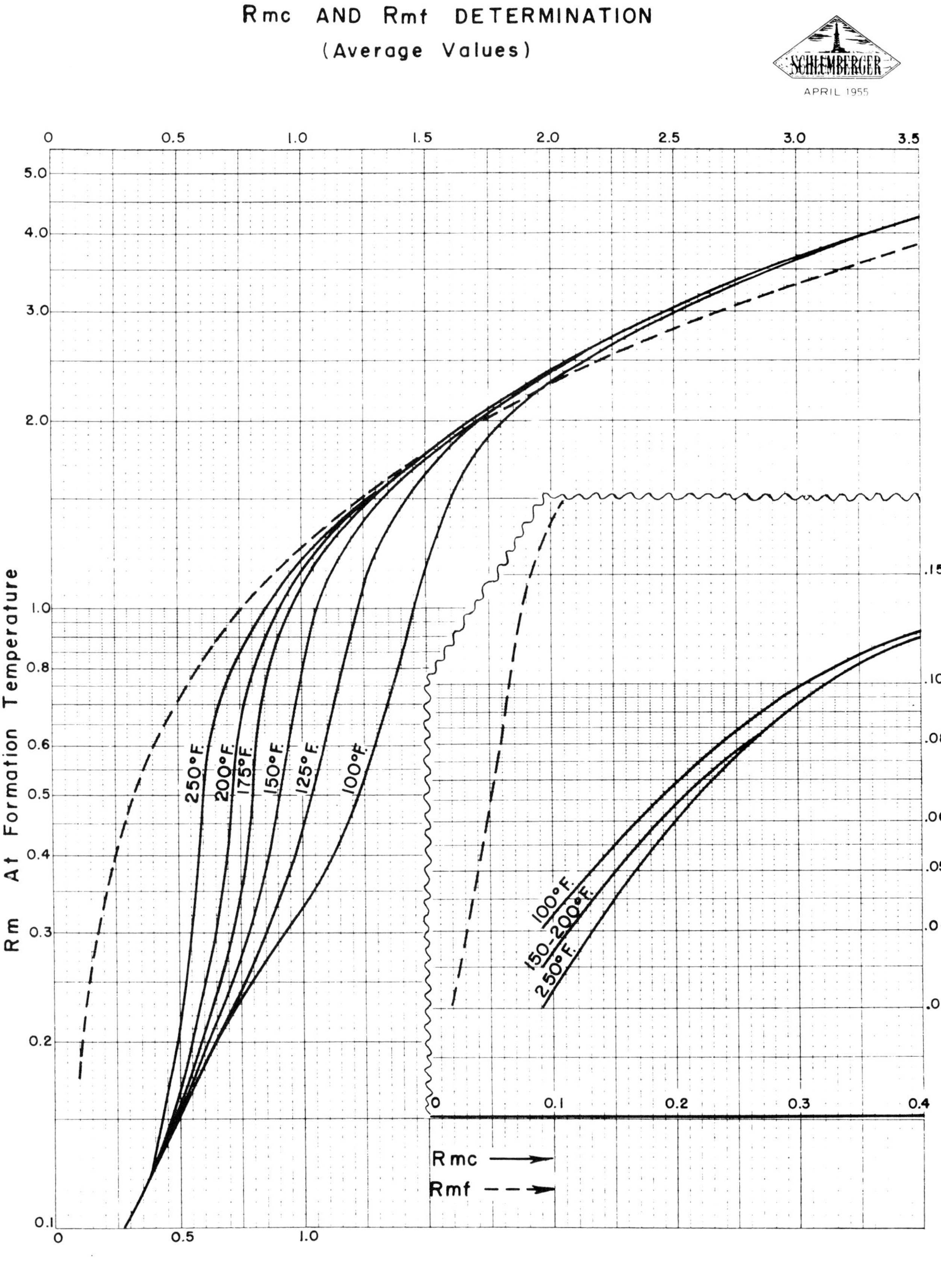
Rmc AND Rmf DETERMINATION
(Average Values)
SCHLUMBERGER
APRIL 1955
0
0.5
1.0
1.5
2.0
2.5
3.0
3.5
5.0
4.0
3.0
2.0
1.0
0.8
0.6
0.5
0.4
0.3
0.2
0.1
Rm At Formation Temperature
250°F.
200°F.
175°F.
150°F.
125°F.
100°F.
.15
Rm
.10
.08
.06
.05
.04
.03
100°F.
150-200°F.
250°F.
0
0.1
0.2
0.3
0.4
Rmc
Rmf
0
0.5
1.0

Chart A-4

S C H L U M B E R G E R   W E L L   S U R V E Y I N G   C O R P.

# LOG INTERPRETATION CHARTS

## RESISTIVITY — SALINITY CHART

### *USE AND DESCRIPTION*

This chart is used:

1. To convert the resistivity of a fluid at one temperature to its resistivity at another temperature.

2. To convert the resistivity of a water sample to the NaCl concentration in parts per million or grains per gallon.

3. To convert concentration of a water sample of a solution to resistivity at a given temperature.

Data on salinity of NaCl solutions related to temperature and to resistivity are plotted, being derived from the International Critical Tables.

### *INFORMATION REQUIRED*

$R_m$, $R_{mf}$, and $R_w$ at a Known Temperature and Formation Temperature

or

Equivalent NaCl Salinity and Temperature

or

Chemical Analysis of the Solution and Temperature

### *PROCEDURE*

*If used to adjust resistivity due to a change in temperature:* Plot a point corresponding to a measured resistivity and temperature. Enter a point for the measured fluid resistivity (read resistivity on lower horizontal scale) and its temperature (read temperature on left vertical scale). Move this point parallel to the diagonal lines (which represent constant salinity values) to the new temperature and then read on the bottom scale the new resistivity value.

*If used to find $R_w$ from a chemical analysis of a solution:* Convert the solution to terms of equivalent NaCl by applying appropriate multipliers* for the separate ions, then finding the sum. Enter this value at the top and follow the diagonal line to the appropriate temperature and then read the resistivity at the bottom. If water analysis is in terms of chloride only, multiply this concentration by the factor 1.65 to find NaCl concentration. Note that the NaCl concentration is in terms of p.p.m., parts per million or micrograms/gram of solution, or in terms of G/G, grains per gallon of solution. Number of G/G × 17.1 = number of p.p.m.

| * MULTIPLIERS: | | |
|---|---|---|
| | Na = 1.0 | Cl = 1.0 |
| | Ca = 0.95 | $SO_4$ = 0.5 |
| | Mg = 2.0 | $CO_3$ = 1.26 |
| | | $HCO_3$ = 0.27 |

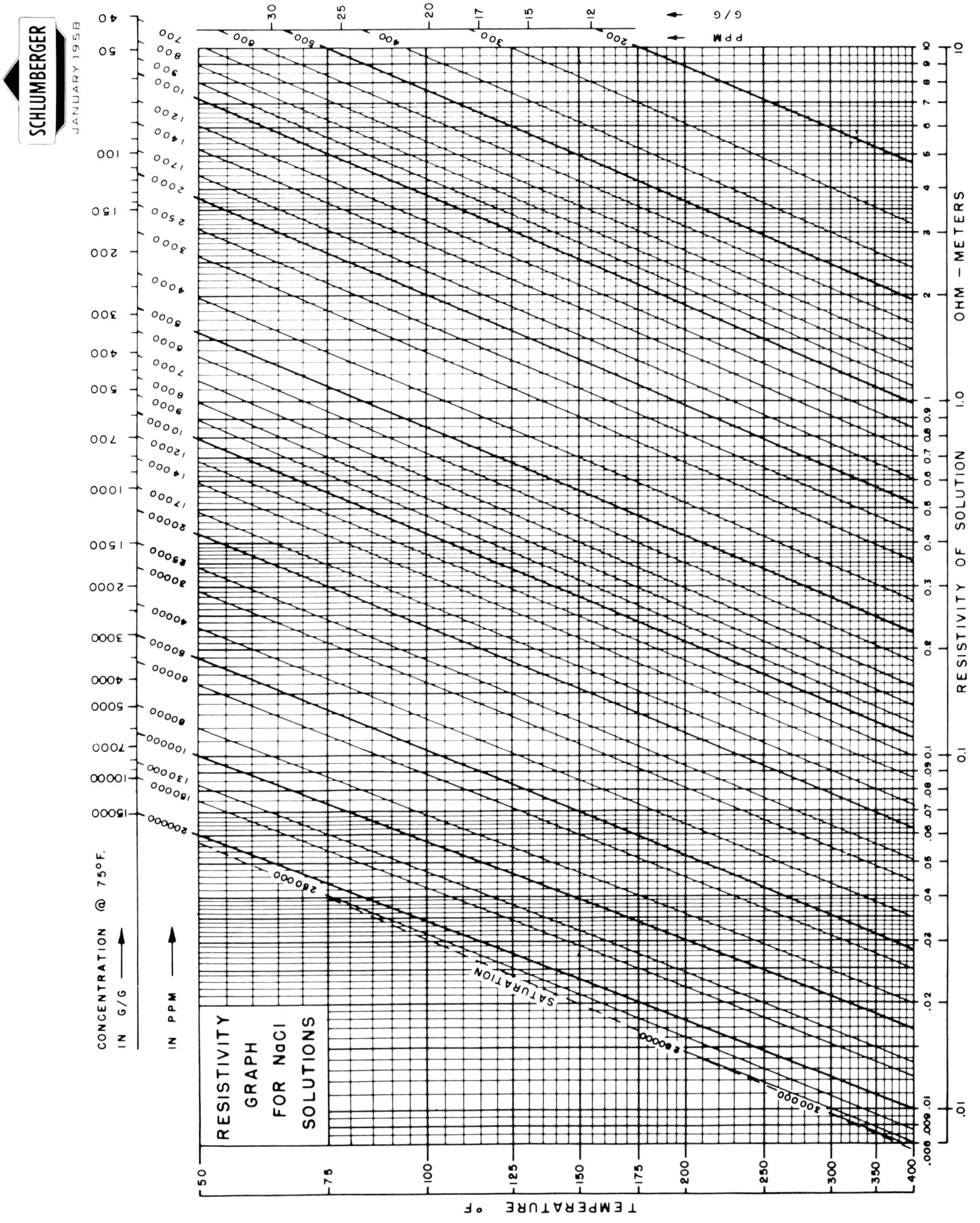
SCHLUMBERGER
JANUARY 1958
RESISTIVITY GRAPH FOR NaCl SOLUTIONS
CONCENTRATION @ 75°F.
IN G/G
IN PPM
G/G
PPM
SATURATION
RESISTIVITY OF SOLUTION OHM – METERS
TEMPERATURE °F

Chart A-6

SCHLUMBERGER WELL SURVEYING CORP.

# LOG INTERPRETATION CHARTS

## SP CORRECTION CHART

*USE AND DESCRIPTION*

This provisional chart is used to correct the observed SP when it is too low because of bed thickness. The effect of invasion is also accounted for. The data used to construct this chart are largely empirical, and inaccuracies may be present when $R_i/R_m$ is high. This applies to average hole sizes of about 8″ or 9″.

*INFORMATION REQUIRED*

Bed Thickness with help of SP Inflection Point or MicroLog

$R_i/R_m$ from 16″ Normal—See Chart B-10.

*PROCEDURE*

Enter the bed thickness at the bottom; proceed up to an appropriate curve of $R_i/R_m$; then move to the left where a correction factor (multiplier) is given.

*EXAMPLE*

GIVEN: $R_i/R_m = 50$.

Bed Thickness = 10′.

Observed SP = –60 MV.

SOLUTION: Correction Factor is 1.33.

Corrected SP = –80 MV.

SPECIAL CASE: The curve of $R_i/R_m = 5$ is used for low resistivity formations.

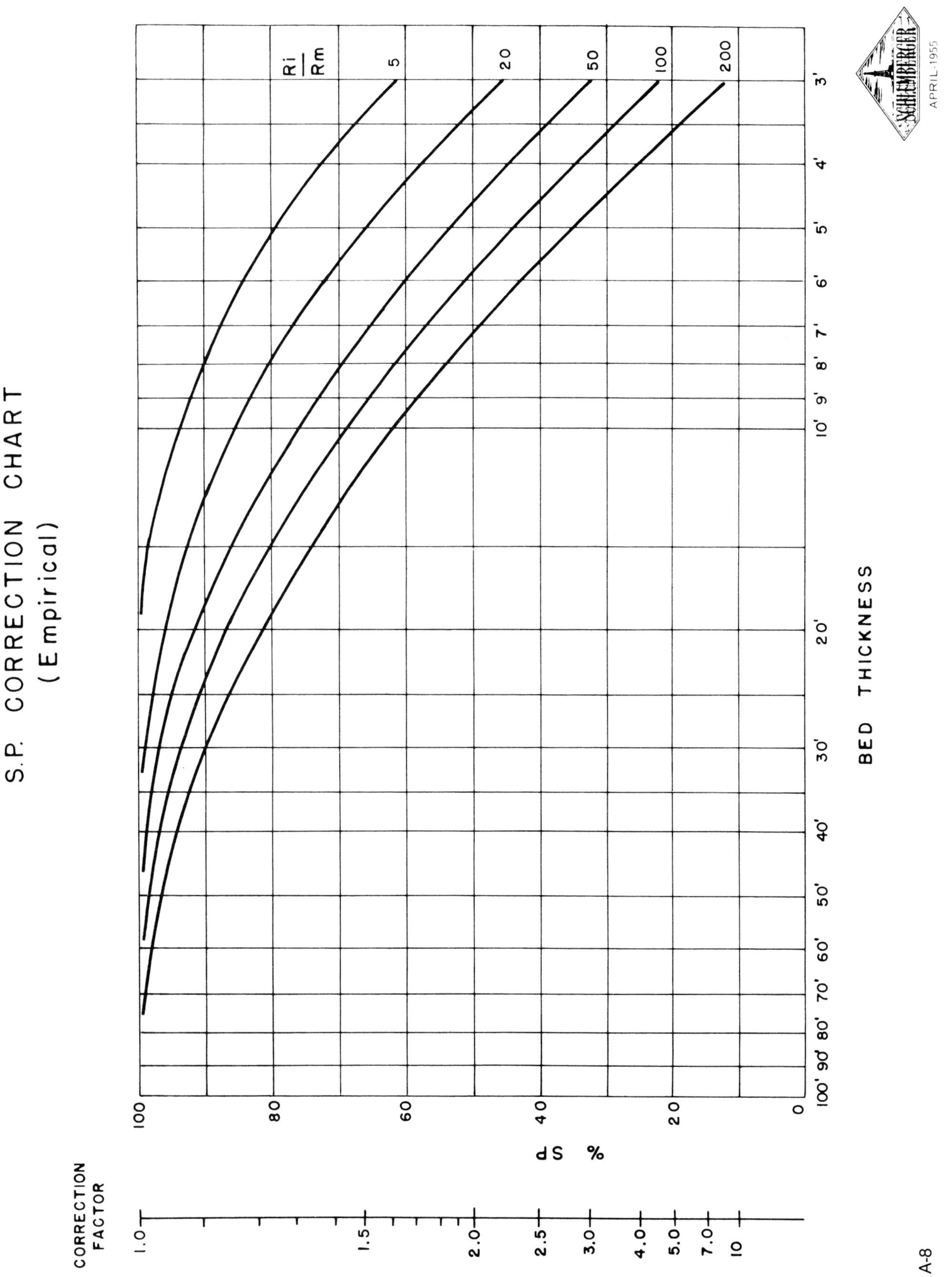
S.P. CORRECTION CHART
(Empirical)
Ri/Rm
5
20
50
100
200
3'
4'
5'
6'
7'
8'
9'
10'
20'
30'
40'
50'
60'
70'
80'
90'
100'
BED THICKNESS
100
80
60
40
20
0
% SP
CORRECTION FACTOR
1.0
1.5
2.0
2.5
3.0
4.0
5.0
7.0
10
SCHLUMBERGER
APRIL-1955

Chart A-8

SCHLUMBERGER WELL SURVEYING CORP.

## LOG INTERPRETATION CHARTS

## $R_w$ DETERMINATION FROM THE SP

### PART I. FINDING $(R_w)_e$

*USE AND DESCRIPTION*

The electrochemical component of the SP, $E_c$, is correctly related to the activity of the formation water and of the mud filtrate, such that $E_c = -K \log a_w/a_{mf}$, where K varies with temperature.

Also a relation such as $E_c = -K \log R_{mf}/(R_w)_e$ can be used. The term $(R_w)_e$, the equivalent water resistivity, is often nearly equal to the true $R_w$; but, when the formation waters are either very saline or quite fresh, important deviations occur.*

Therefore, the first step is to find $(R_w)_e$ from the static SP reading. The second step is to convert $(R_w)_e$ to $R_w$ by using Chart A-12.

*SPECIAL CASES: SALINE MUDS*

If the mud filtrate, $R_{mf}$, is less than 0.1 ohm-m at 75°F, the effect of its increased activity must also be considered. The relation now becomes $E_c = -K \log (R_{mf})_e/(R_w)_e$. An extra step is required to convert $R_{mf}$ to $(R_{mf})_e$ by Chart A-12.

*MUDS CONTAINING GYPSUM OR $CaCl_2$*

For muds containing Ca or Mg in solution, special procedures must be used (see reference below). Lime base muds usually have a negligible amount of Ca in solution and may be treated as regular mud types.

*INFORMATION REQUIRED*

SP, Corrected for bed thickness if necessary (Chart A-8)
$R_{mf}$ at a known temperature
Formation Temperature (Chart A-2) *or* Knowledge of K

*PROCEDURE*

The SP is entered; proceed up to either a line equivalent to the formation temperature or to a line representing K. On left margin read $R_{mf}/(R_w)_e$. This is then divided into the $R_{mf}$ at formation temperature, giving $(R_w)_e$.

*EXAMPLE*

*Fresh Mud*

GIVEN: SP = −100 MV at 190°F.
$R_{mf}$ = 0.45 ohm-m at 190°F.

SOLUTION: $R_{mf}/(R_w)_e = 15.$ (K = − 85.)

$$(R_w)_e = \frac{R_{mf}}{R_{mf}/(R_w)_e} = 0.45/15.$$

$(R_w)_e = .03.$

*Saline Mud*

SP = +40 MV at 150°F.
$R_{mf}$ = .025 ohm-m at 150°F.

$R_{mf}$ corresponds to $(R_{mf})_e = .013.$ (A-12)

$$(R_w)_e = \frac{(R_{mf})_e}{(R_{mf})_e/(R_w)_e} = .013/.32 = .04.$$

$(R_{mf})_e/(R_w)_e = 0.32.$

*Gondouin, M., Tixier, M. P., and Simard, G. L. "An Experimental Study on the Influence of the Chemical Composition of Electrolytes on the SP Curve." Preprint, American Institute of Mining, Metallurgical and Petroleum Engineers Meetings: Casper, Wyoming, May, 1956 and Los Angeles, California, October, 1956.

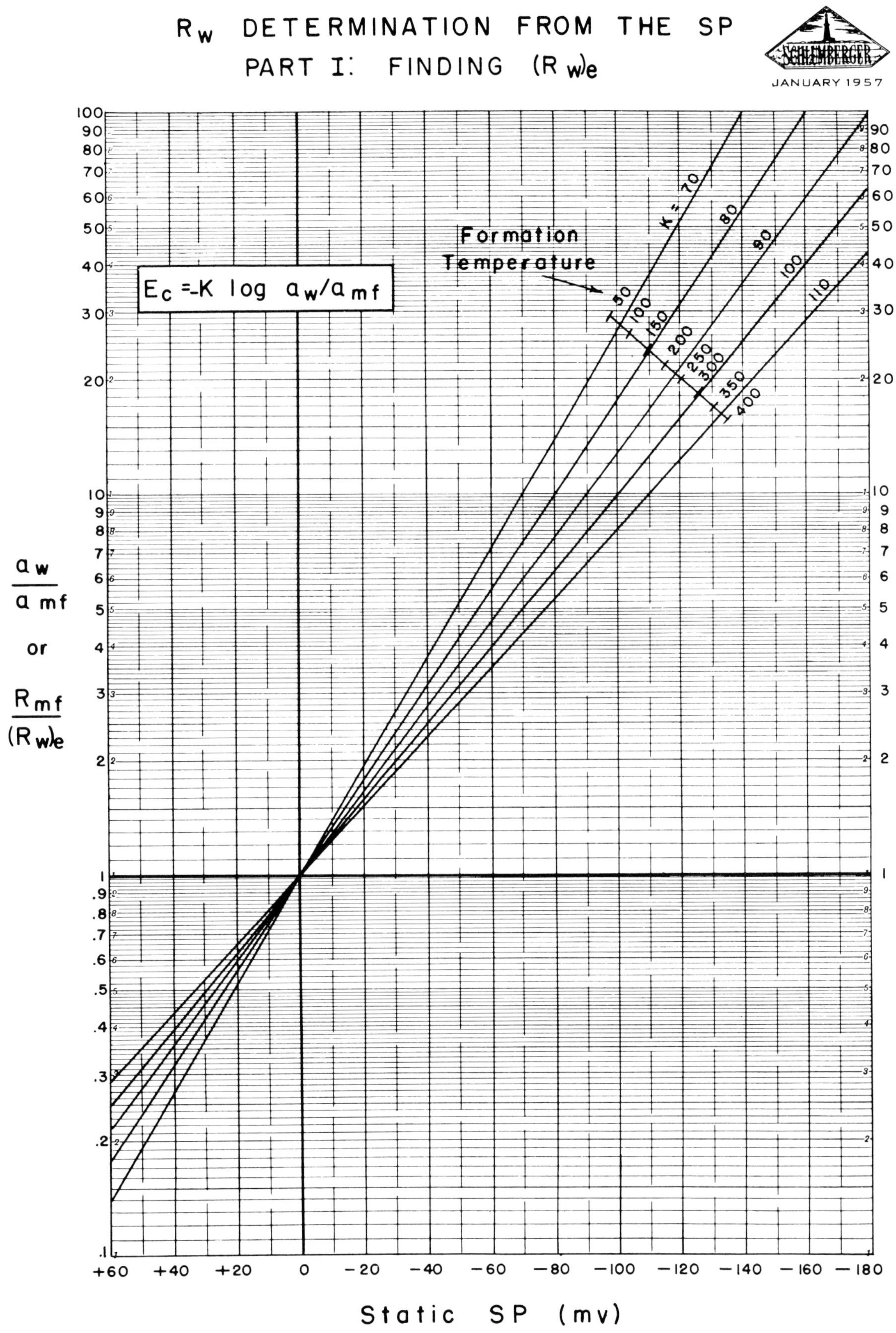
Rw DETERMINATION FROM THE SP
PART I: FINDING (Rw)e
SCHLUMBERGER
JANUARY 1957
Ec = -K log aw/amf
Formation Temperature
K = 70
80
90
100
110
50
100
150
200
250
300
350
400
aw / amf or Rmf / (Rw)e
100
90
80
70
60
50
40
30
20
10
9
8
7
6
5
4
3
2
1
.9
.8
.7
.6
.5
.4
.3
.2
.1
+60
+40
+20
0
−20
−40
−60
−80
−100
−120
−140
−160
−180
Static SP (mv)

Chart A-10

SCHLUMBERGER WELL SURVEYING CORP.

# LOG INTERPRETATION CHARTS

## $R_w$ DETERMINATION FROM THE SP

### PART II. TRUE $R_w$ VS. EQUIVALENT $(R_w)_e$

*USE AND DESCRIPTION*

This chart provides a means of correcting the equivalent $(R_w)_e$ to the true $R_w$ (at BHT). It can be separated into two parts:

A. Where $R_w > 0.1$ at BHT

These curves are plotted for average fresh formation waters (Where Ca and Mg become important) as reported in the Gondouin, Tixier, and Simard paper.* Local experience and chemical analysis of the waters may show these curves somewhat in error in special cases.

B. Where $R_w < 0.1$ at BHT.

These curves are constructed from activity considerations and are theoretically correct.

*INFORMATION REQUIRED*

$(R_w)_e$ at a specified temperature (from Chart A-10).

*PROCEDURE*

Enter $(R_w)_e$ at the bottom, go up to the curve for the correct temperature, and then to the left to read $R_w$ at that temperature.

*EXAMPLE:*

*Fresh Mud*

From Page A-9, we found:
$(R_w)_e = .03$ at 190°F.
$R_w = .037$ ohm-m at 190°F.

*Saline Mud*

$(R_w)_e = .04$ at 150°F.
$R_w = .05$ ohm-m at 150°F.

*Gondouin, M., Tixier, M. P., and Simard, G. L. "An Experimental Study on the Influence of the Chemical Composition of Electrolytes on the SP Curve." Preprint, American Institute of Mining, Metallurgical and Petroleum Engineers Meetings: Casper, Wyoming, May, 1956 and Los Angeles, California, October, 1956.

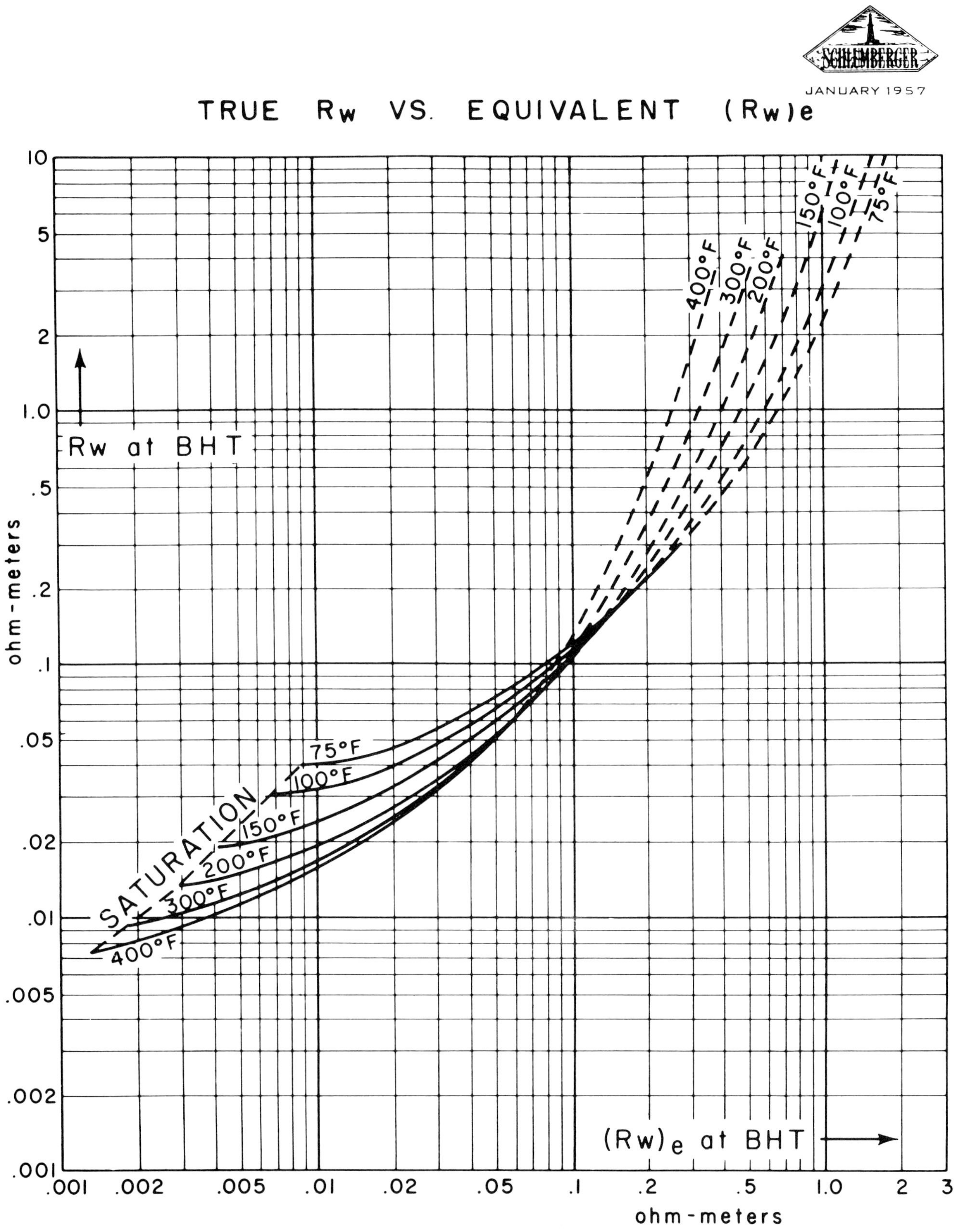
SCHLUMBERGER
JANUARY 1957
TRUE Rw VS. EQUIVALENT (Rw)e
Rw at BHT
ohm-meters
(Rw)e at BHT
ohm-meters
SATURATION
75°F
100°F
150°F
200°F
300°F
400°F
10
5
2
1.0
.5
.2
.1
.05
.02
.01
.005
.002
.001
.001 .002 .005 .01 .02 .05 .1 .2 .5 1.0 2 3

Chart A-12

SCHLUMBERGER WELL SURVEYING CORP.

# LOG INTERPRETATION CHARTS

## SIMPLIFIED RESISTIVITY DEPARTURE CURVES

### *USE AND DESCRIPTION*

This chart corrects for the influence of the bore hole on the 16″ normal and for the hole and average invasion on the 18′8″ lateral. The sonde diameter and the effect of the reference electrodes are taken into account.

It is used in relatively thick formations of moderate and high resistivity where the 16″ normal is influenced by the invaded zone.

### *INFORMATION REQUIRED*

$R_m$ at Formation Temperature

Actual Hole Diameter

Apparent 16″ and 18′8″ Resistivities

### *PROCEDURE*

*For the 16″ normal:* Find the ratio $R_{16''}/R_m$. Enter at the left margin and proceed to the solid curve for appropriate hole size. Read $R_i/R_m$ at the bottom.

*For the 18′8″ lateral:* Find the ratio $R_{18'8''}/R_m$. Enter at the left margin and proceed to the dashed line for appropriate hole size and read $R_t/R_m$ at the bottom.

### *EXAMPLE*

GIVEN: $R_m = .5$ ohm at formation temperature.

Hole Size = 9″.

$R_{16''} = 50.$

$R_{18'8''} = 250.$

SOLUTION: $R_{16''}/R_m = 100.$

$R_{18'8''}/R_m = 500.$

$R_i/R_m = 150.$

$R_t/R_m = 300.$

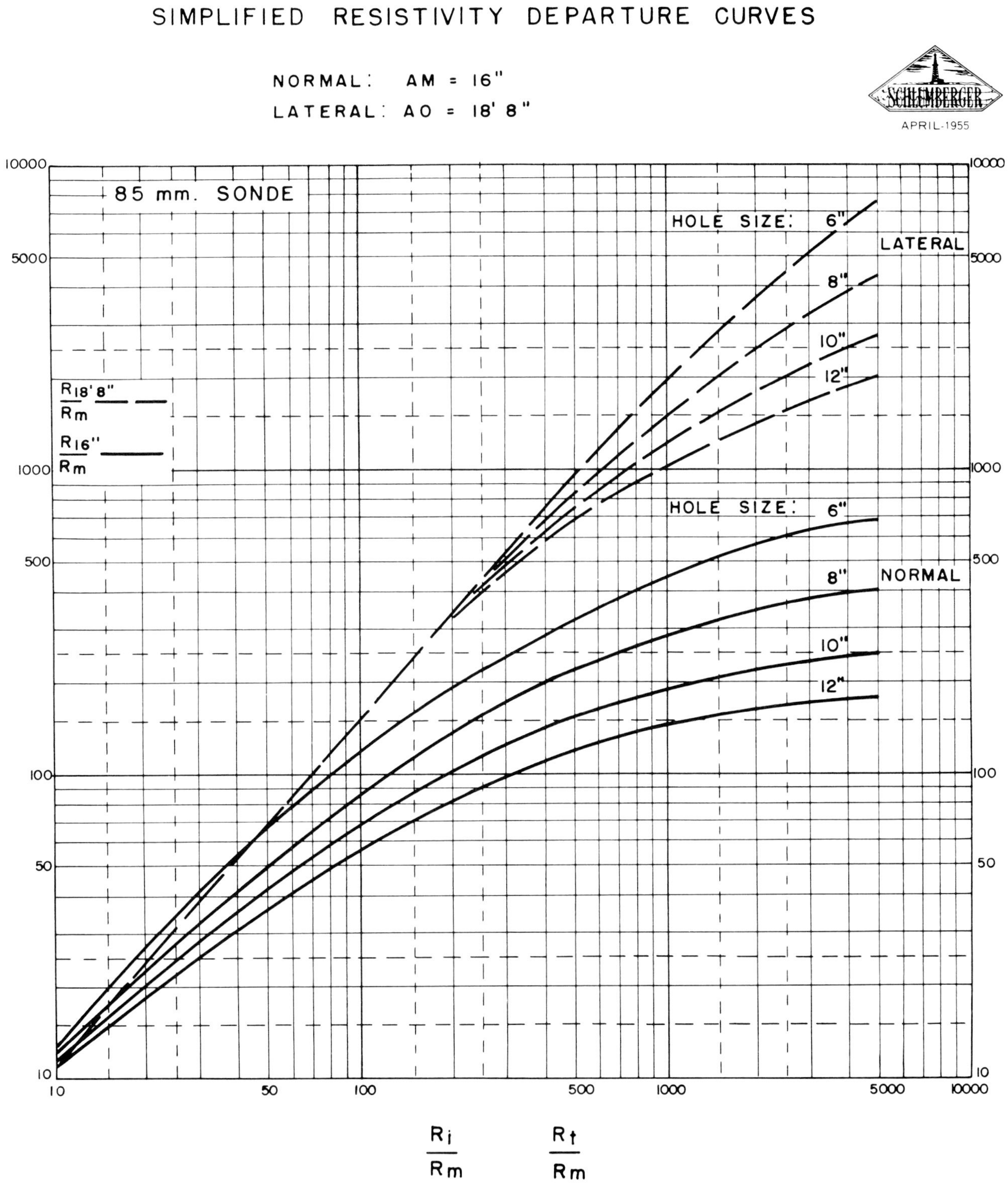
SIMPLIFIED RESISTIVITY DEPARTURE CURVES
NORMAL: AM = 16"
LATERAL: AO = 18' 8"
SCHLUMBERGER
APRIL-1955
85 mm. SONDE
HOLE SIZE: 6"
LATERAL
8"
10"
12"
R18'8" / Rm
R16" / Rm
HOLE SIZE: 6"
8"
NORMAL
10"
12"
10000
5000
1000
500
100
50
10
10
50
100
500
1000
5000
10000
Ri / Rm
Rt / Rm

Chart B-2

SCHLUMBERGER WELL SURVEYING CORP.

# LOG INTERPRETATION CHARTS

## SIMPLIFIED RESISTIVITY DEPARTURE CURVES — PERMIAN BASIN

*USE AND DESCRIPTION*

This chart corrects for the influence of the bore hole on the 10″ normal and for the hole and average invasion on the 18′8″ lateral. The sonde diameter and the effect of the reference electrodes are taken into account.

It is used in relativly thick formations of moderate and high resistivity where the 10″ normal is influenced by the invaded zone.

*INFORMATION REQUIRED*

$R_m$ at Formation Temperature

Actual Hole Diameter

Apparent 10″ and 18′8″ Resistivities

*PROCEDURE*

*For the 10″ normal:* Find the ratio $R_{10''}/R_m$. Enter at the left margin and proceed to the solid curve for appropriate hole size. Read $R_i/R_m$ at the bottom.

*For the 18′8″ lateral:* Find the ratio $R_{18'8''}/R_m$. Enter at the left margin and proceed to the dashed line for appropriate hole size and read $R_t/R_m$ at the bottom.

*EXAMPLE*

GIVEN: $R_m = .5$ ohm at formation temperature.

Hole Size = 9″.

$R_{10''} = 50.$

$R_{18'8''} = 250.$

SOLUTION: $R_{10''}/R_m = 100.$

$R_{18'8''}/R_m = 500.$

$R_i/R_m = 250.$

$R_t/R_m = 300.$

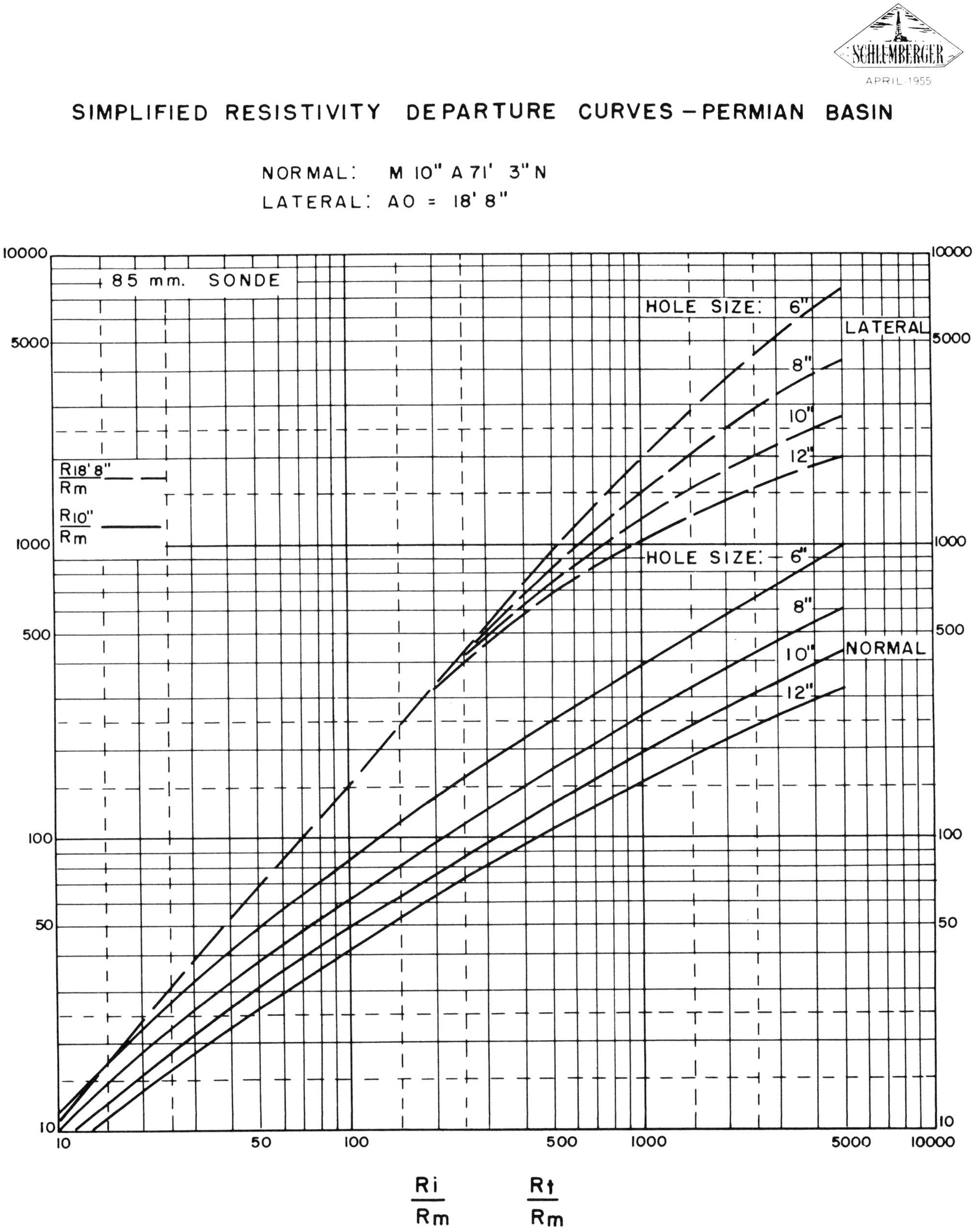
SCHLUMBERGER
APRIL 1955
SIMPLIFIED RESISTIVITY DEPARTURE CURVES — PERMIAN BASIN
NORMAL: M 10" A 71' 3" N
LATERAL: AO = 18' 8"
85 mm. SONDE
R18'8" / Rm
R10" / Rm
HOLE SIZE: 6"
8"
10"
12"
LATERAL
HOLE SIZE: 6"
8"
10"
12"
NORMAL
10000
5000
1000
500
100
50
10
10
50
100
500
1000
5000
10000
Ri / Rm
Rt / Rm

Chart B-4

## $R_t$ ESTIMATION FROM ELECTRICAL LOGS

Selection of the proper logging method is very important. Recommended programs, based upon formation and mud type, are given on page 2 of this chartbook.

The Electrical Log is at its best when used in fresh muds, formations of low to moderate resistivity, and fairly thick beds. Departure from these conditions complicates determination of $R_t$.

Rules for obtaining $R_t$ from Electrical Logs are based on the relative resistivities of the beds compared to the resistivities of the mud and surrounding formations. Therefore, formations are subdivided into three classes, depending on the ratio $R_{16''}/R_m$. These simplifying rules are derived from a study of resistivity departure curves.

1) Low Resistivity — when $R_{16''}/R_m < 10$.

The shorter spacings, such as 16″ and 64″ normals are most useful in finding $R_t$. Often $R_m \simeq R_s$, in which case the apparent value of the 64″ normal can be easily adjusted to $R_t$, depending on the ratio $R_{64''}/R_s$ and the bed thickness. The upper part of page B-8 gives this information.

Correction of the 16″ normal for thin beds is shown on Charts B-12 and B-13.

2) Medium Resistivity — when $10 < R_{16''}/R_m < 50$.

In this case the 64″ normal is very useful in the lower portion of this resistivity range, but when $R_{16''}/R_m >$ 20, use of the 18′8″ lateral becomes important, either to find $R_t$, or to confirm the apparent 64″ normal value. The lateral, being an unsymmetrical curve, requires picking of $R_t$ as is shown on page B-8, *Rules for using Lateral.*

3) High Resistivity — When $R_{16''}/R_m > 50$.

The 64″ normal is greatly affected by invasion, so that the 18′8″ lateral is the best choice for estimating $R_t$. Choice of the values to use are as shown on page B-8, but these must be further corrected by Chart B-2 (or B-4). In some very tight formations invasion may be such that $R_t$ cannot be found by any device, in which case $R_i$ and $R_t$ as found from Chart B-2 (or B-4) tend to approach the same value. Also, at this time the ratio of this apparent $R_t$ compared to a reconstructed $R_o$ tends to approach the ratio $R_{mf}/R_w$ in nonproductive, slightly permeable formations.

Application of the Porosity Balance is desirable if an independent porosity is available. This method will help to confirm the $R_t$ value (Refer to page D-13).

*Laterolog:* The Laterolog is particularly useful in this type of resistivity, mainly because of the detailed response obtained and the elimination of a borehole effect. As the equations at the bottom of page B-8 show, invasion affects the Laterolog considerably; but the corrections are simple. Required are a measurement of $R_{xo}$ and a selection of $D_i$, which must be determined by experience. Invasion effects are minimized when $R_{xo}$ is equal to or less than $R_t$, a condition met with salt muds.

SCHLUMBERGER
OCTOBER 1958

# $R_t$ ESTIMATION FROM ELECTRICAL LOGS

| BED THICKNESS (e) | QUALIFICATIONS | | DEVICE | RESPONSE |
|---|---|---|---|---|
| **A. IN LOW RESISTIVITY, WHEN $R_{16''}/R_m < 10$ (INVASION UP TO 2d)** | | | | |
| e > 20' (> 4 AM') | | | Long Normal | $R_{64''} = R_t$ |
| $e \simeq 15'$ ( 3 AM') | $R_m \simeq R_s$ | $R_{64''}/R_s \geq 2.5$ | Long Normal | $R_{64''} = 2/3\ R_t$ |
| $e \simeq 15'$ ( 3 AM') | $R_m \simeq R_s$ | $R_{64''}/R_s \leq 1.5$ | Long Normal | $R_{64''} = R_t$ |
| $e \simeq 10'$ ( 2 AM') | $R_m \simeq R_s$ | $R_{64''}/R_s \geq 2.5$ | Long Normal | $R_{64''} = 1/2\ R_t$ |
| $e \simeq 10'$ ( 2 AM') | $R_m \simeq R_s$ | $R_{64''}/R_s = 1.5$ | Long Normal | $R_{64''} = 2/3\ R_t$ |
| 5' < e < 10' | When oil bearing and SP is -50 – 80 MV | | Short Normal | $R_{16''} \simeq R_t$ |
| 5' < e < 10' | Surrounding beds homogenous | | Lateral in resistive bed | $R_t \gtrsim R_{Max} \times R_s/R_{Min}$ |
| Thin beds (in general) | Surrounding beds homogenous | | Lateral in conductive bed | $R_{19''} \simeq R_t$ |

B. RULES FOR USING LATERAL (AO = 18'8")

e > 40' ( > 2.0 AO)

Use Mid-point Method

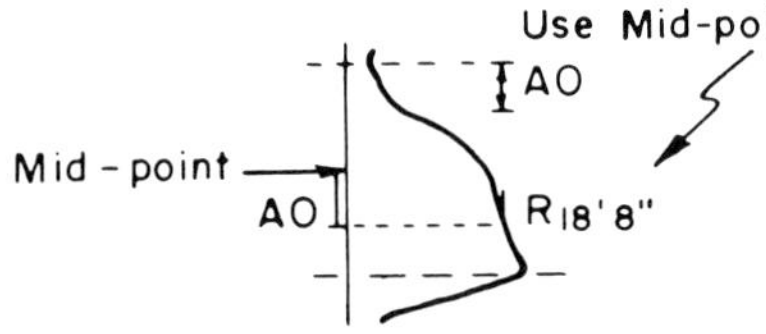


$e \simeq 28'$ ( = 1.5 AO )

Use 2/3 Rule

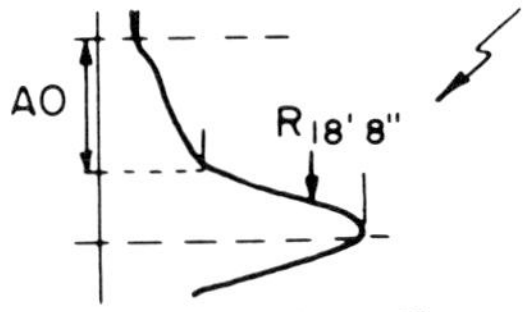


$e \simeq 24'$ ( = 1.3 AO)

Use $R_{Max}$

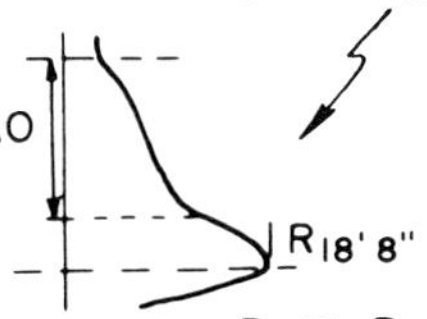


5' < e < 10' — Resistive bed and surrounding beds homogeneous

$R_t \gtrsim R_{Max} \times R_s/R_{Min}$

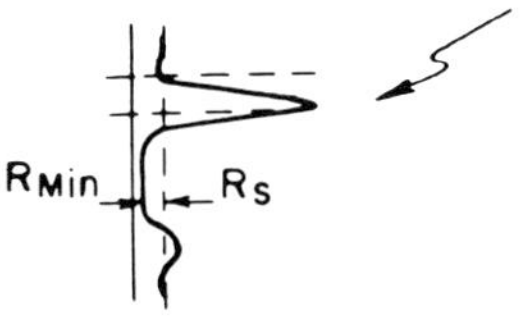


(When $R_{16''}/R_m > 50$, these values must then be corrected for the bore hole.) (Chart B-2)

C. RESPONSE OF LATEROLOG – 7 AND LATEROLOG – 3

| | | |
|---|---|---|
| For Di = 20" | $R_{LL7} \simeq 0.2\ R_{xo} + 0.8\ R_t$ | $R_{LL3} \simeq 0.25\ R_{xo} + 0.75\ R_t$ |
| Di = 40" | $R_{LL7} \simeq 0.4\ R_{xo} + 0.6\ R_t$ | $R_{LL3} \simeq 0.5\ R_{xo} + 0.5\ R_t$ |
| Di = 80" | $R_{LL7} \simeq 0.6\ R_{xo} + 0.4\ R_t$ | $R_{LL3} \simeq 0.75\ R_{xo} + 0.25\ R_t$ |

Thus, the best results occur when $R_{xo} < R_t$ and $R_{mf}/R_w < 4$

Chart B-8

SCHLUMBERGER WELL SURVEYING CORPORATION

# 16'' NORMAL CORRECTION CHART
# OLD E-LOGS

## USE AND DESCRIPTION

This chart will correct the 16'' Normal resistivity for thin beds. The corrected value is the value which a 16'' Normal would read in a very thick bed with the same invasion.

### *INFORMATION REQUIRED*

e - bed thickness
$R_{16''}$ - reading at the center of the bed
$R_s$ - surrounding bed resistivity
$R_m$ - mud resistivity at formation temperature

### *PROCEDURE*

Determine $R_{16''}/R_m$

Enter this value at the appropriate hole size chart on the left margin; go to the proper bed thickness curve and then to the bottom to find the $R_{16''}/R_m$ corrected. Multiply by $R_m$ to get $R_{16''}$corrected.

### *EXAMPLE*

Given: $R_{16''} = 5$
$R_s = 1$
$R_m = 1.25$
e = 3'
d = 8'

Solution: $R_{16''}/R_m = 4$

Therefore, $R_{16''}/R_m$ corrected = 8
$R_{16''}$ corrected = 8 x 1.25 = 10

16" NORMAL CORRECTION CHART – THIN BEDS

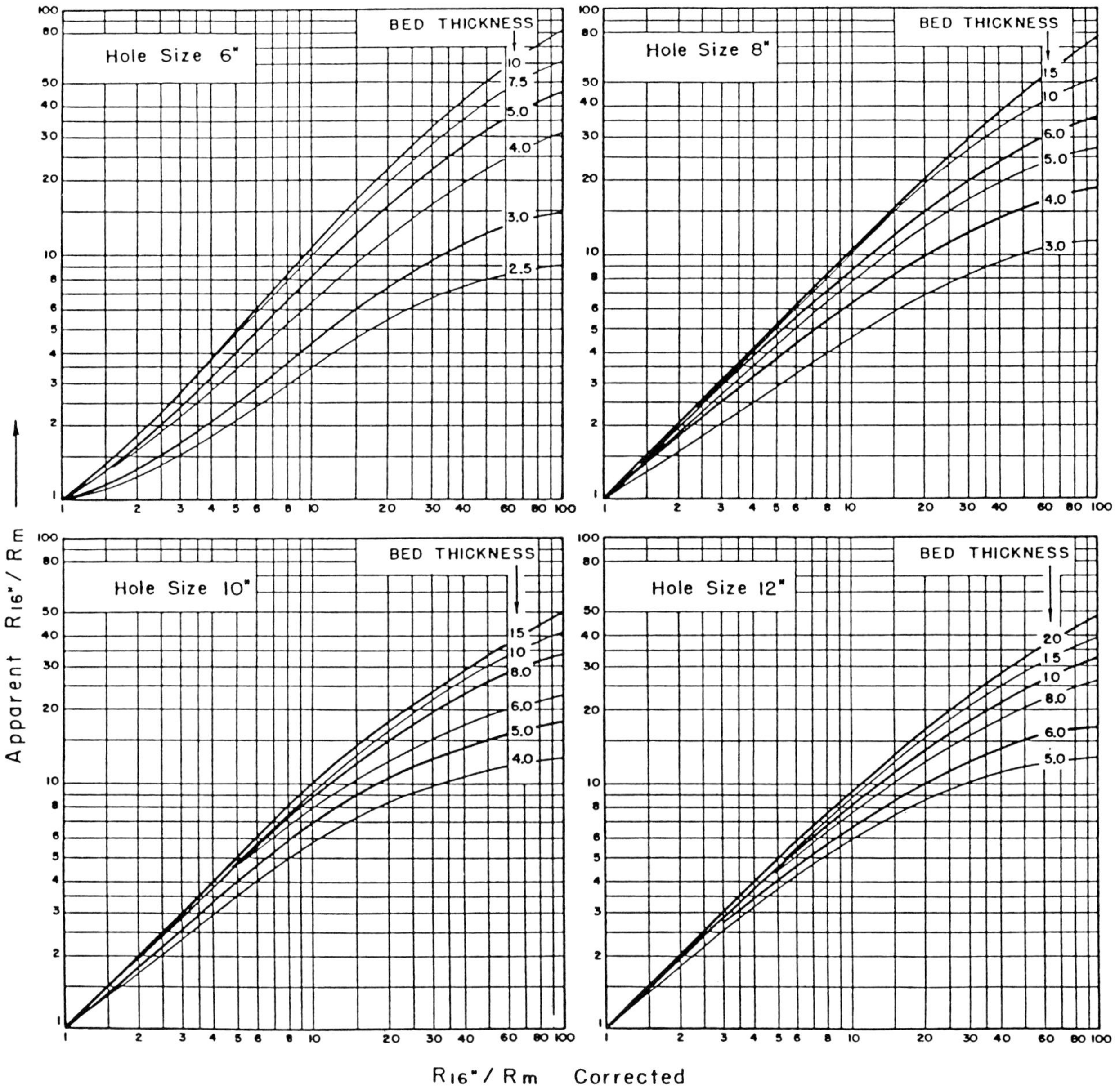


Conditions for use : When $R_m \simeq R_s$ , or when $R_t > 7 R_s$ ( for $e > 6'$ )
or when $R_t > 10 R_s$ ( for $e < 6'$ )

Chart B-10

S C H L U M B E R G E R   W E L L   S U R V E Y I N G   C O R P.

# LOG INTERPRETATION CHARTS

## EMPIRICAL MICROLOG CHART

*USE AND DESCRIPTION*

This chart is used to find porosity and formation factor by use of the MicroLog. It is an empirical chart statistically built from comparisons between core analysis and the micronormal (2″) as read from Type D pad MicroLogs. Therefore, average mud cake thickness and residual oil are assumed. Thicker mud cakes make the porosity indicated as too large, and thinner mud cakes the porosity too small.

Use with the hydraulic type pad will give porosities which tend to be too low.

At the bottom of the sheet is attached an alignment graph giving the relation between porosity and formation factor, using the Humble Formula.

*INFORMATION REQUIRED*

$R_m$ at Formation Temperature

$R_{2''}$ read from the Dotted Curve of the MicroLog

*PROCEDURE*

Enter $R_m$ at the left and $R_{2''}$ at the bottom. Intersection of lines from these points permits estimation of porosity. Formation factor can also be noted, but for best interpolation the estimated porosity should be referred to the alignment graph at the bottom.

*EXAMPLE*

GIVEN: $R_m = 0.7$ at Formation Temperature.
$R_{2''} = 6$.

SOLUTION: Porosity = 12%.
Formation Factor = 62.

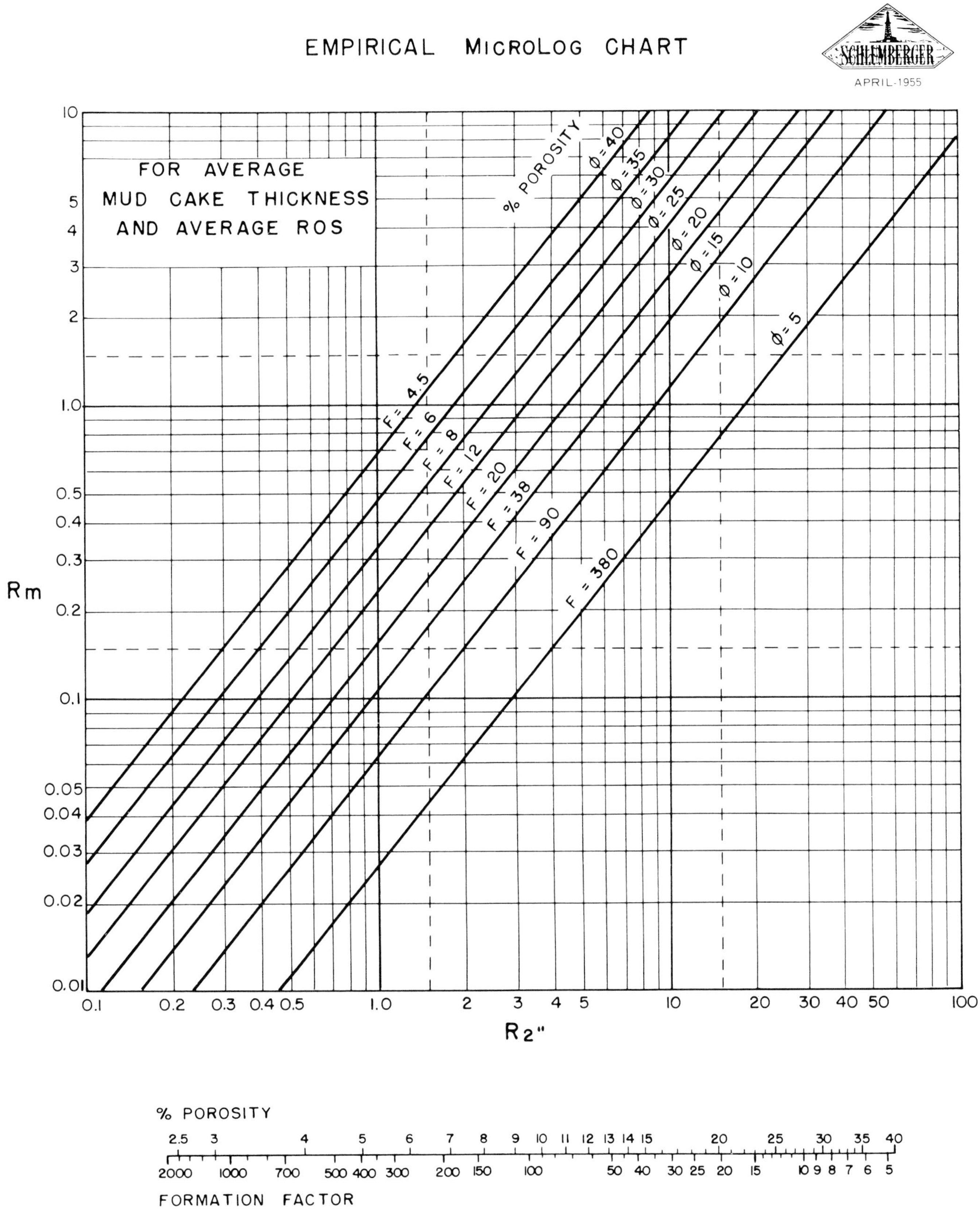
EMPIRICAL MICROLOG CHART
SCHLUMBERGER
APRIL-1955
FOR AVERAGE MUD CAKE THICKNESS AND AVERAGE ROS
% POROSITY
φ = 40
φ = 35
φ = 30
φ = 25
φ = 20
φ = 15
φ = 10
φ = 5
F = 4.5
F = 6
F = 8
F = 12
F = 20
F = 38
F = 90
F = 380
Rm
10
5
4
3
2
1.0
0.5
0.4
0.3
0.2
0.1
0.05
0.04
0.03
0.02
0.01
R2"
0.1 0.2 0.3 0.4 0.5 1.0 2 3 4 5 10 20 30 40 50 100
% POROSITY
2.5 3 4 5 6 7 8 9 10 11 12 13 14 15 20 25 30 35 40
2000 1000 700 500 400 300 200 150 100 50 40 30 25 20 15 10 9 8 7 6 5
FORMATION FACTOR

Chart C-2

SCHLUMBERGER WELL SURVEYING CORP.

# LOG INTERPRETATION CHARTS

## SERIES C — INTERPRETATION CHART

**(For Hydraulic Pads on Noninsulated Sonde in 8″ Hole)**

*USE AND DESCRIPTION*

This chart has been built by laboratory measurements and relates the response of the MicroLog curves to both the mudcake thickness and to the flushed zone resistivity, $R_{xo}$. All these terms are expressed with $R_{mc}$ as a base for measurement. This is the fundamental chart for MicroLogs *WITH B AT BACK PLATE,* and provides the information used to construct the nomographic Chart C-8. The reliability is best when $t_{mc} < ½''$ and $R_{xo}/R_{mc} < 25$.

*INFORMATION REQUIRED*

$R_{mc}$ — Mud cake resistivity at formation temperature

$R_{1''x1''}$ — Value of the microinverse

$R_{2''}$ — Value of the micronormal

d — Approximate hole size

*PROCEDURE*

Obtain an average $R_{1''x1''}$ and $R_{2''}$ from the MicroLog. Divide each by $R_{mc}$ (measured, or from Chart A-4). Enter these on their respective axes to establish a point within the chart. Refer to the dashed lines to estimate $t_{mc}$, the mud cake thickness on the formation. Refer to the solid lines to estimate $R_{xo}/R_{mc}$. Go to Chart C-12 to find $R_{xo}$, $\phi$, F.

*EXAMPLES:*

GIVEN: $R_{mc} = 1.5$

$R_{1''x1''} = 4.5$

$R_2 = 5.0$

$R_{1''x1''}/R_{mc} = 3.0$; $R_{2''}/R_{mc} = 3.33$

SOLUTION: $t_{mc} = 3/16''$

$R_{xo}/R_{mc} = 6.0$

$R_{xo} = 9.0$ (for 8″ hole)

NOTE: For holes other than 8″ multiply $R_{1''x1''}/R_{mc}$ by these factors before entering the chart:

4¾″ . . . . . . . . . . . . 1.15

6 ″ . . . . . . . . . . . . 1.05

10 ″ . . . . . . . . . . . . 0.93

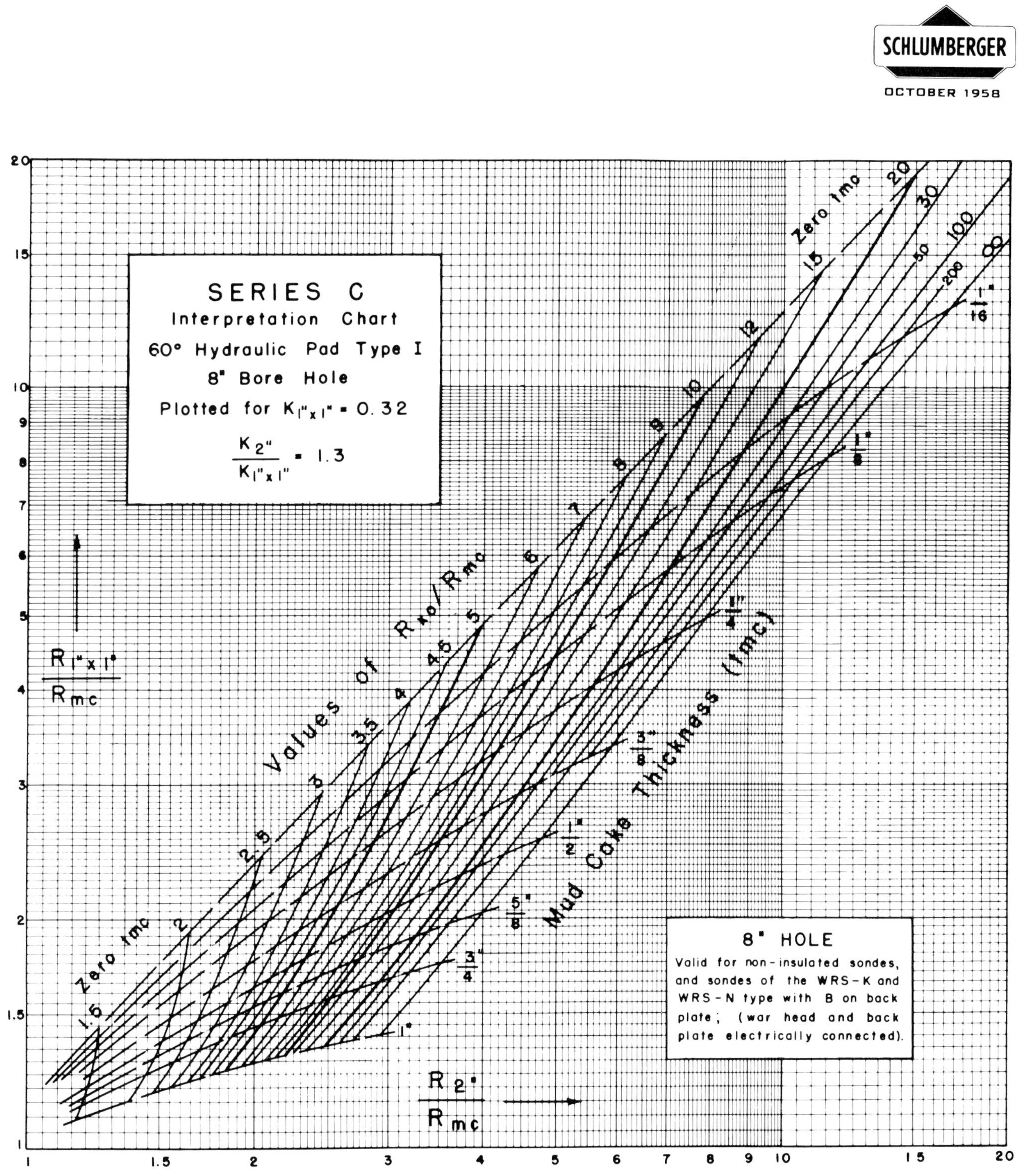


NOTE: An incorrect $R_{mc}$ will displace the points in the chart along a 45° line passing through that point. In certain cases this can be recognized when the mud cake thickness is different from direct measurement by the microcaliper. To correct, move the point at 45° to intersect the known $t_{mc}$. Read $R_{xo}/R_{mc'}$ at the middle and $R_{2''}/R_{mc'}$ at the bottom of the chart. Divide the log value of $R_{2''}$ by the ratio $R_{2''}/R_{mc'}$ to find $R_{mc'}$. Now, $R_{xo}/R_{mc'}$ times $R_{mc'}$ equals $R_{xo}$.

Chart C-4

SCHLUMBERGER WELL SURVEYING CORP.

## $R_{xo}$ DETERMINATION ($R_{mc}$ UNKNOWN)

*USE AND DESCRIPTION*

By replotting Chart C-4, the term $R_{mc}$ can be eliminated. In place of $R_{mc}$, a good value of mud cake thickness ($t_{mc}$) is required. However, the accuracy in $t_{mc}$ is less important if $t_{mc} > \frac{1}{2}''$. Therefore, the chart is particularly recommended for thick mud cakes.

*INFORMATION REQUIRED*

$R_{1''x1''}$ and $R_{2''}$ — from hydraulic pad tool, "B" at backplate.

d — approximate hole size.

$t_{mc}$ — thickness of the mud cake.

*PROCEDURE*

Obtain the ratio of the microcurves: $R_{1''x1''}/R_{2''}$. Enter on the left for the closest hole size. Project to the appropriate mud cake thickness curve. Drop vertically to read $R_{xo}/R_{2''}$. This ratio is multiplied by $R_{2''}$ to give $R_{xo}$.

*EXAMPLE*

Given: $d = 10''$.

$t_{mc} = \frac{3}{4}''$.

$R_{1''x1''} = 3.5$.

$R_{2''} = 5$.

Solution: $R_{1''x1''}/R_{2''} = .7$, computed.

$R_{xo}/R_{2''} = 2.8$, from chart.

$R_{xo} = 2.8 \times 5 = 14$.

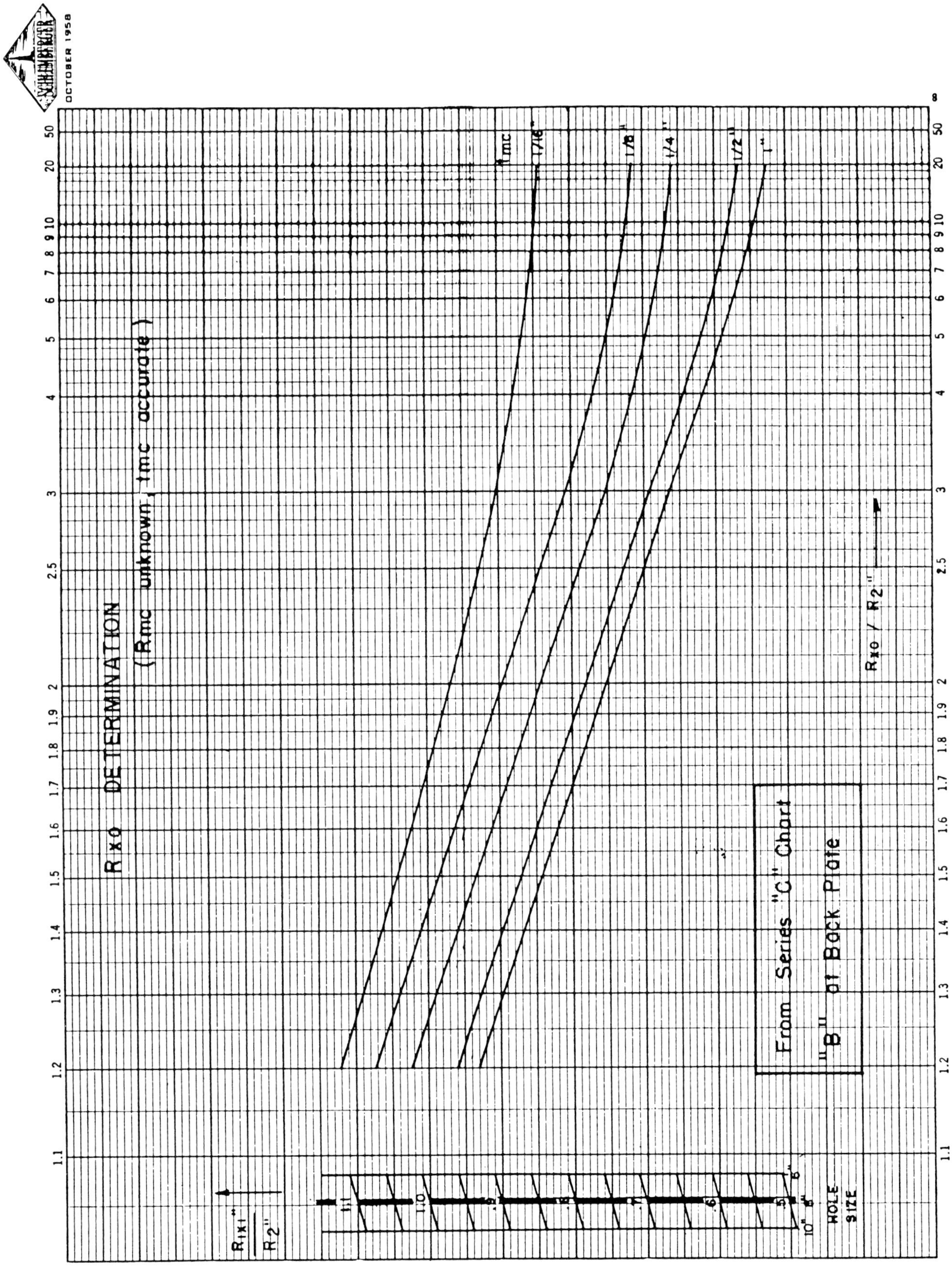
SCHLUMBERGER
OCTOBER 1958
Rxo DETERMINATION
(Rmc unknown, tmc accurate)
From Series "C" Chart
"B" at Back Plate
Rxo / R2"
R1x1" / R2"
tmc
1/16"
1/8"
1/4"
1/2"
1"
HOLE SIZE

Chart C-6

SCHLUMBERGER WELL SURVEYING CORP.

# LOG INTERPRETATION CHARTS

## MICROLOG NOMOGRAPHIC CHART — "D" PAD

*USE AND DESCRIPTION*

This chart is used in the determination of $R_{xo}$ which is essential for formation factor determination and for use in the shaly sand method of interpretation. It is a nomographic solution for $R_{xo}/R_{mc}$ from "D" pad MicroLog. Note that accuracy is poor when $R_{16''}/R_m > 10$ unless the mud cake thickness on the formation is of the order of ⅛″ or less and when the value found for $R_{xo}/R_{mc}$ is less than 25. Thus, it is useful mostly in soft rock regions.

*INFORMATION REQUIRED*

$R_{mc}$ at Formation Temperature — measured, or from Chart A-4

$R_{1''x1''}$ from Microinverse of MicroLog

$R_{2''}$ from Micronormal of MicroLog

Hole Size (d)

*PROCEDURE*

Find the ratios $R_{1''x1''}/R_{mc}$ and $R_{2''}/R_{mc}$. Plot these points on their respective scales. Place a straightedge through these points and extend to heavy $R_{xo}/R_{mc}$ scale line which applies to an 8″ hole. For larger holes, a correction to larger values is required when $R_{xo}/R_{mc} > 3$; to correct, proceed from the 8″ hole scale horizontally to the correct hole size and estimate the $R_{xo}/R_{mc}$ from the relative position between lines of equal $R_{xo}/R_{mc}$ values.

For a 6″ hole, the correction is made by multiplying the $R_{1''x1''}/R_{mc}$ value by 1.07, then proceeding as for an 8″ hole.

*EXAMPLE*

GIVEN: $R_{mc} = 1.0$.

$R_{1''x1''} = 4.5$.

$R_{2''} = 5.0$.

$d = 10''$.

SOLUTION: $R_{xo}/R_{mc} = 7.0$, for 8″ hole.

Corrected $R_{xo}/R_{mc} = 7.8$, for 10″ hole.

$R_{xo} = 7.8$.

## NOMOGRAPHIC "MicroLog" CHART – "D" PAD

APRIL 1955

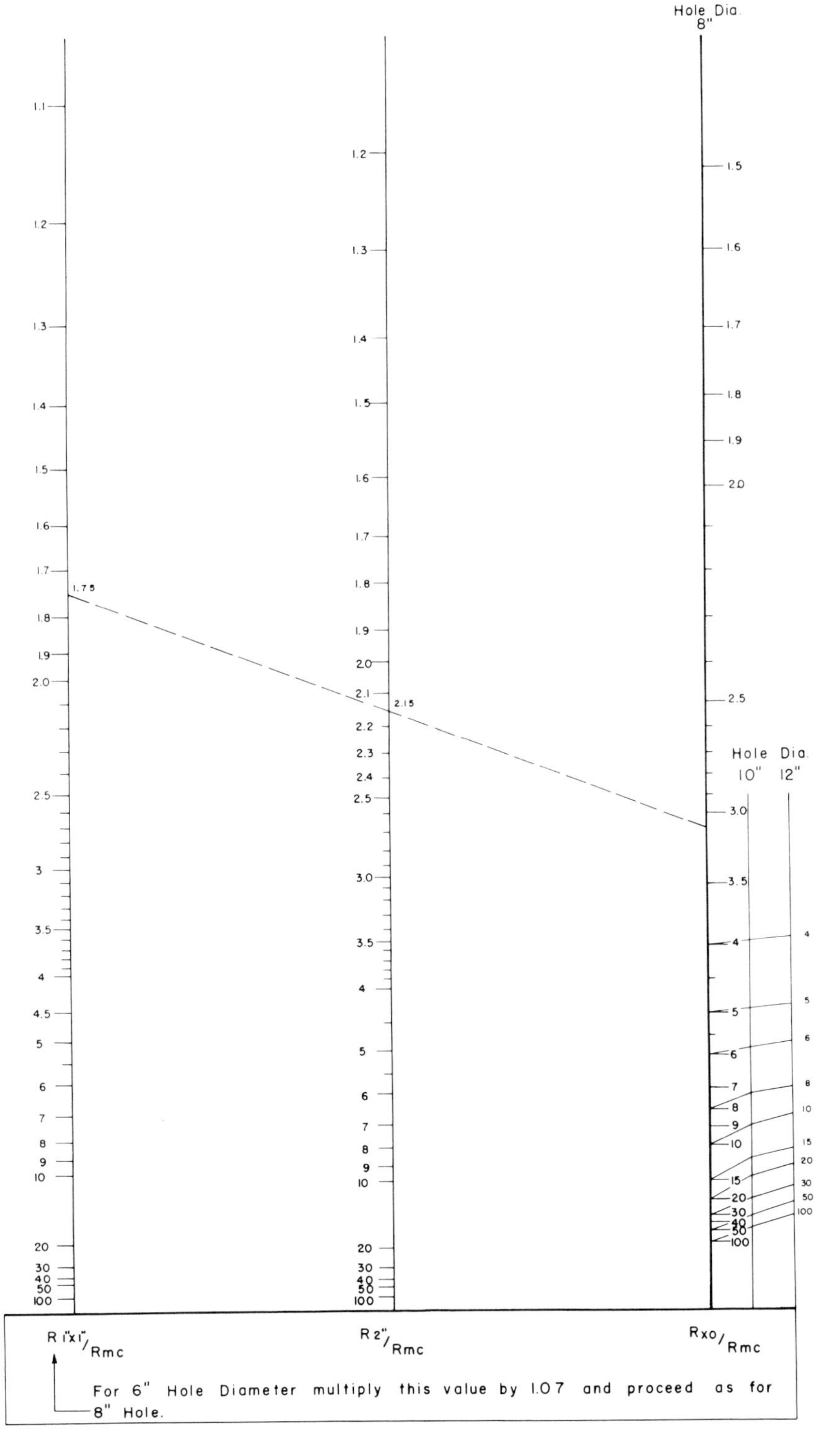


To be used only when R16"/Rm is less than 10, & above this range when the mud cake thickness equals or is less than 1/8".

Chart C-8

# SCHLUMBERGER WELL SURVEYING CORPORATION

## FORMATION FACTOR–POROSITY

*USE AND DESCRIPTION*

This chart is used to convert porosity to formation factor or formation factor to porosity. Two basic formulas are employed. The Humble equation

$$F = \frac{0.62}{\phi^{2.15}}$$

recommended for use in sands and sandstones; and the Archie equation

$$F = \frac{1}{\phi^{m}}$$

recommended, with appropriate cementation factor, for use in carbonates. Several values of m - cementation factor are shown.

*INFORMATION REQUIRED*

$\phi$ – porosity *or*

F – formation factor

Proper formation factor-porosity relationship.

*PROCEDURE*

Determination of formation factor: Enter the porosity in ordinate and go horizontally to the appropriate formation factor-porosity relationship. From this intercept, go vertically to the abscissa to read formation factor.

Determination of porosity: Enter the formation factor in abscissa and go vertically to the appropriate formation factor-porosity relationship. From this intercept, go horizontally to the ordinate to read porosity.

*EXAMPLES*

GIVEN: $\phi$ = 29%
Sandstone

SOLUTION: F = 9 (Humble equation)

GIVEN: F = 65
Carbonate (m = 2.2)

SOLUTION: $\phi$ = 15%

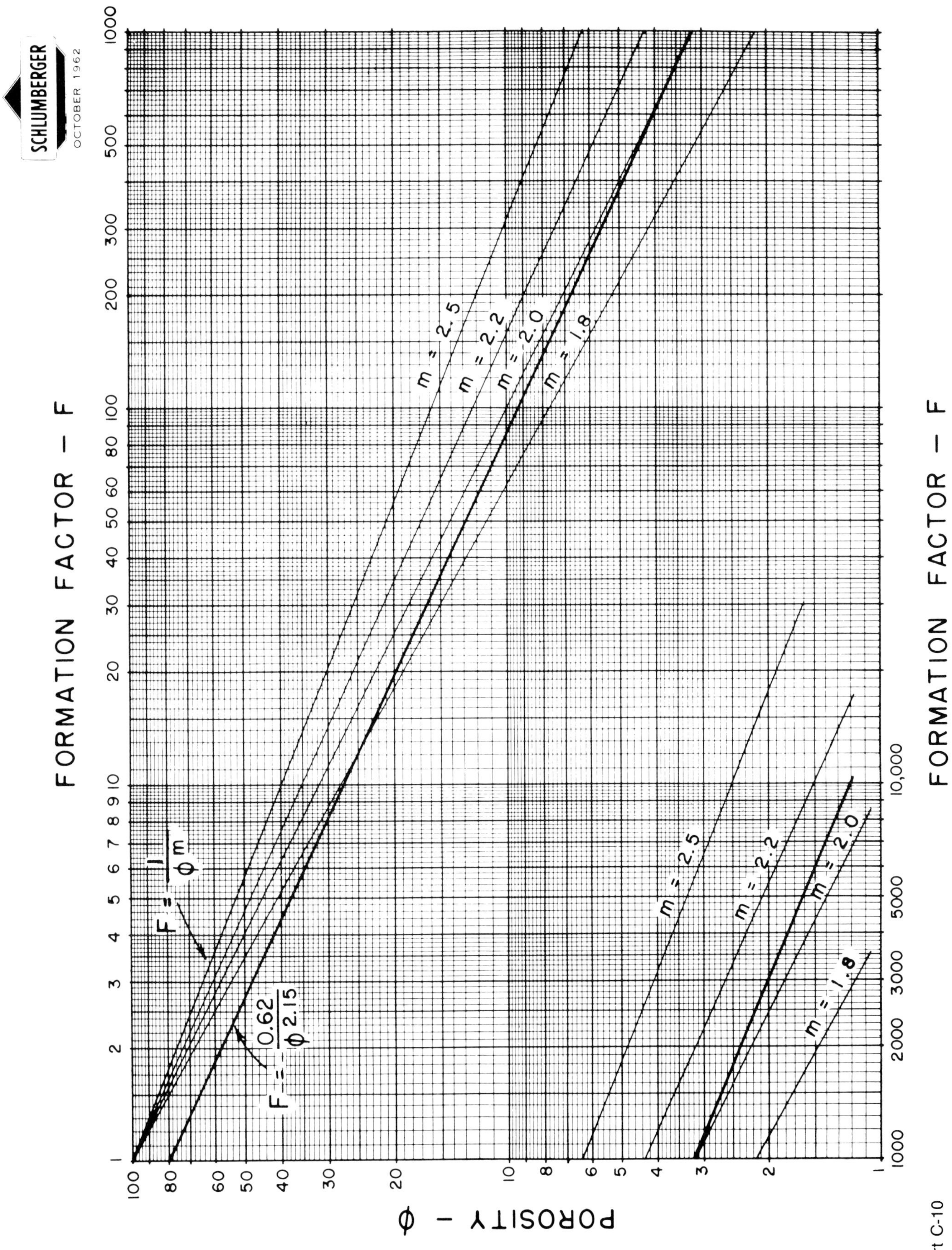
SCHLUMBERGER
OCTOBER 1962
FORMATION FACTOR — F
POROSITY — φ
m = 2.5
m = 2.2
m = 2.0
m = 1.8
F = 1/φ^m
F = 0.62/φ^2.15


Chart C-10

## SCHLUMBERGER WELL SURVEYING CORPORATION

### POROSITY AND FORMATION FACTOR DETERMINATION

*USE AND DESCRIPTION*

This chart takes the $R_{xo}/R_{mc}$ values found from Charts C-4 and C-14 and computes $R_{xo}$, F, and $\phi$ by nomographic procedures. It provides a means of correcting for Residual Oil Saturation (ROS).

Porosity ($\phi$) is found by either the Humble or Archie Relation.

*INFORMATION REQUIRED*

$R_{xo}/R_{mc}$ – from Charts C-4 or C-14
$R_{mc}$ – from mud cake resistivity already used in above determination
$R_{mf}$ – measured mud filtrate or from Chart A-4, at formation temperature
ROS – assumed
$m$ – Cementation exponent (for Archie porosity relation)
*Special Case*
$R_{xo}$ – read directly from Microlaterolog when mud cake is thin

*PROCEDURE*

Enter $R_{xo}/R_{mc}$ at the left; pass a line through the $R_{mc}$ value to find $R_{xo}$. Now, divide $R_{xo}$ by $R_{mf}$ by passing through the proper value of $R_{mf}$. This gives $F_a$, which in clean formations is different from F only by the effect of residual oil. Next, proceed horizontally until the assumed value of per cent ROS is reached (scale at top). Then follow the trend of the sloping line to 0% ROS, upon which line is entered the per cent porosity, $\phi$. Equivalent F, the formation factor, is read directly opposite to the right, being related by the Humble formula. If the Archie relation ($F = 1/\phi^m$) is preferred, proceed from this value of equivalent F through the proper $m$ to obtain the porosity.

In the special case, enter $R_{xo}$ scale with Microlaterolog reading and proceed to the right.

*EXAMPLE I*

GIVEN: $R_{xo}/R_{mc}$ = 10
$R_{mc}$ = .7
$R_{mf}$ = .4
ROS = assumed at 15%

SOLUTION: $R_{xo}$ = 7; therefore, $F_a$ = 17.5
$\phi$ = 24.5 (by Humble Formula), with F = 13
If Archie Relation is used with $m$ = 2.0, $\phi$ = 28%

*EXAMPLE II*

GIVEN: $R_{MLL}$ = 20; $t_{mc} < \frac{1}{4}$"; $R_{mf}$ = 0.05; ROS = 20%
SOLUTION: $R_{xo}$ = 20; $F_a$ = 400
$\phi \simeq$ 6%; F = 260

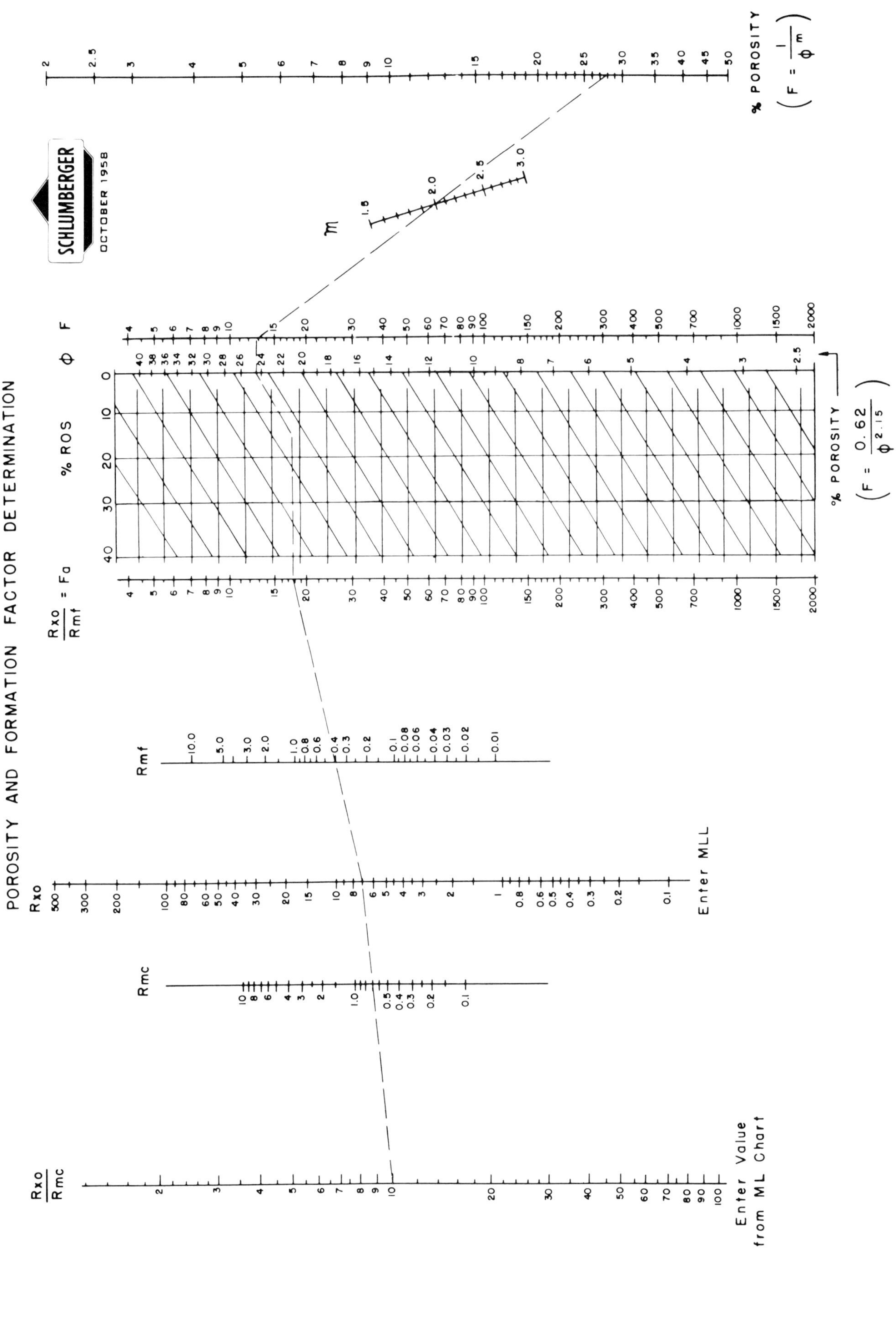
POROSITY AND FORMATION FACTOR DETERMINATION
SCHLUMBERGER
OCTOBER 1958
% POROSITY
$\left(F = \frac{1}{\phi^m}\right)$
m
% ROS
$\phi$
F
$\frac{R_{xo}}{R_{mf}} = F_a$
% POROSITY
$\left(F = \frac{0.62}{\phi^{2.15}}\right)$
$R_{mf}$
$R_{xo}$
Enter MLL
$R_{mc}$
$\frac{R_{xo}}{R_{mc}}$
Enter Value from ML Chart

Chart C-12

SCHLUMBERGER WELL SURVEYING CORPORATION

## SIMPLIFIED MICROLATEROLOG DEPARTURE CURVES

*USE AND DESCRIPTION*

This chart shows the Microlaterolog response as influenced by the mud cake in front of permeable formations. The chart is used to find $R_{xo}/R_{mc}$ and, more important, to understand the conditions when the Microlaterolog reading may be considered to give $R_{xo}$ directly.

*INFORMATION REQUIRED*

$R_{mc}$ – Resistivity of the Mud Cake–Measured or from Chart A-4

$R_{MLL}$ – Apparent Value read on Microlaterolog Curve

$t_{mc}$ – Mud Cake Thickness in front of Permeable Zone. This is estimated from the caliper or study of Microlog curves.

*PROCEDURE*

The ratio $R_{MLL}/R_{mc}$ is determined and entered horizontally to the estimated mud cake thickness. The abscissa of this intercept is $R_{xo}/R_{mc}$. $R_{xo}$ is found by multiplying by $R_{mc}$ or by use of Chart C-12, which also solves for F and $\phi$.

*EXAMPLE*

GIVEN: $R_{MLL} = 28$

$R_{mc} = 0.8$

$t_{mc} = 3/8''$

SOLUTION: $R_{MLL}/R_{mc} = 35$

$R_{xo}/R_{mc} = 42$

$R_{xo} = 42 \times 0.8 = 33.6$

## SIMPLIFIED MICROLATEROLOG DEPARTURE CURVES

SCHLUMBERGER
OCTOBER 1962

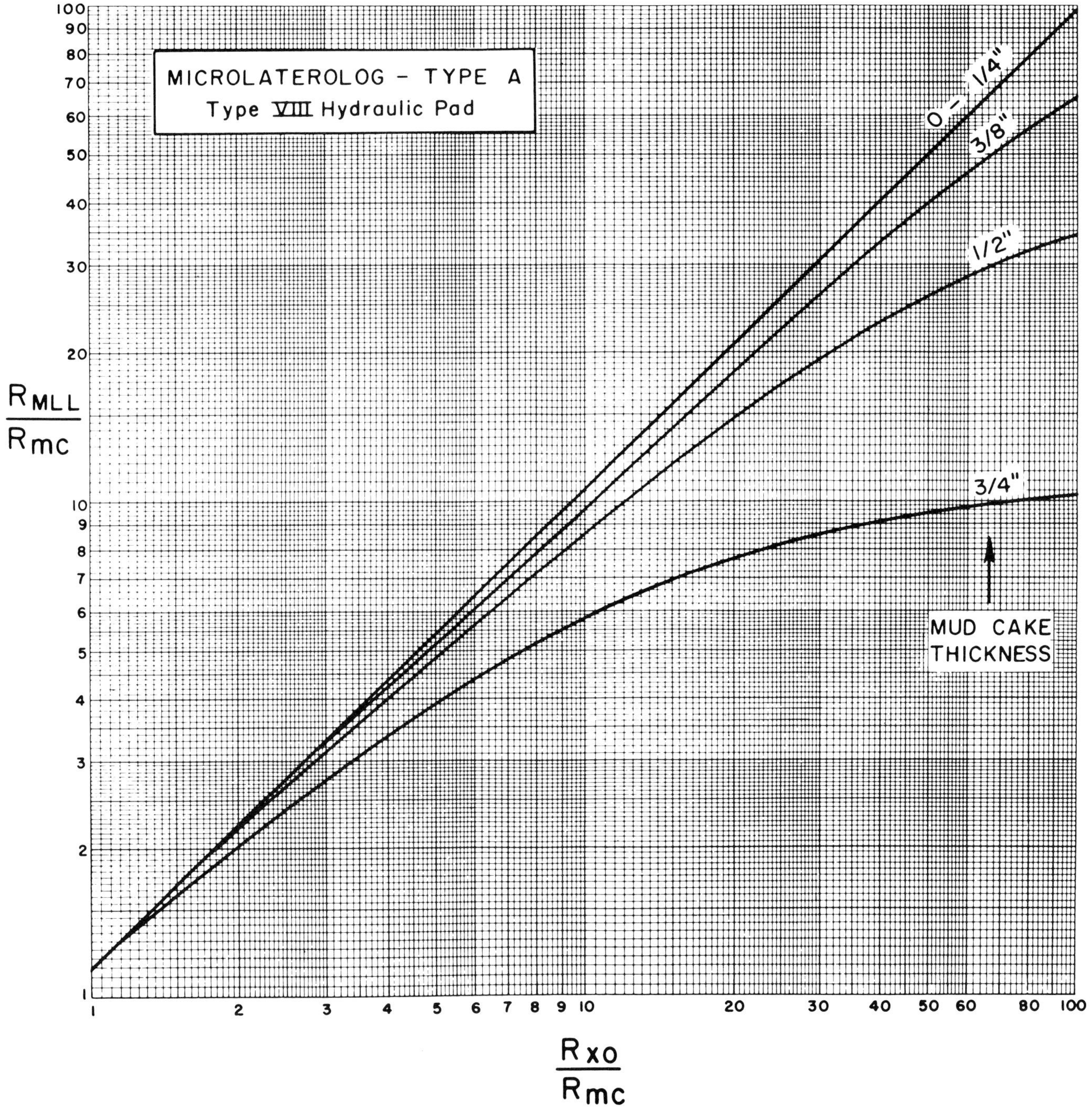


Chart C-14

SCHLUMBERGER WELL SURVEYING CORP.

# LOG INTERPRETATION CHARTS

## POROSITY INDEX FROM LIMESTONE LATERAL

*USE AND DESCRIPTION*

The response of the limestone lateral permits a graphical method for estimating porosity, since the curve levels out at 1±% porosity; the porosity index at any other point is a function of the deflection from zero compared to the maximum deflection for 1±%, providing the hole size is constant and known. This porosity index is then corrected to actual porosity according to the multiplier "C" shown at the bottom right of the chart.

At the bottom left is a table giving maximum $R_a/R_m$ for several bit sizes. This is used to check the $R_m$ value when a zone of 1±% is available. Also, if no such zone is available, the equivalent deflection can be found for 1±% porosity to use for "maximum deflection" at upper left.

*INFORMATION REQUIRED*

$R_{Max}$ — in ohms or in divisions deflection

$R_a$ — the same as above for the zone in question.

Hole Size and AO Spacing

SP

*PROCEDURE*

At the upper left a line is passed through points representing values on the curve for maximum deflection (from a dense zone) and observed deflection to get per cent deflection. This value is entered into the main chart, and intersection with the known hole size gives the porosity index. This value has to be corrected by consideration of the SP, as shown at the bottom right: Porosity = C × Porosity Index.

If no $R_{Max}$ is seen on the log, it is computed by knowing the hole size and multiplying the appropriate maximum $R_a/R_m$ by $R_m$ which is known. See table, lower left.

*EXAMPLE*

GIVEN: $R_a$ = 50 ohms; $R_m$ = .5; SP = —25; Hole Size = 7⅞″; AO = 32″.

SOLUTION: (1). Find maximum deflection. Maximum $R_a/R_m$ for 7⅞″ hole and AO = 32″ = 128 ohms. Then maximum deflection = maximum $R_a/R_m \times R_m$ = 128 × .5 = 64 ohms.

(2). Find per cent deflection. Use 64 on first scale and 50 on second scale; read 80 on third scale.

(3). Find porosity index for 7⅞″ hole. Equals 5.5.

(4). Find porosity — see bottom right for value of "C".
If oil bearing, C = 1.3, since SP = —25.
Porosity = C × PI = 1.3 × 5.5 = 7%.

NOTE: Step (1) would be unnecessary if the maximum deflection in a dense zone was available for the same hole size.

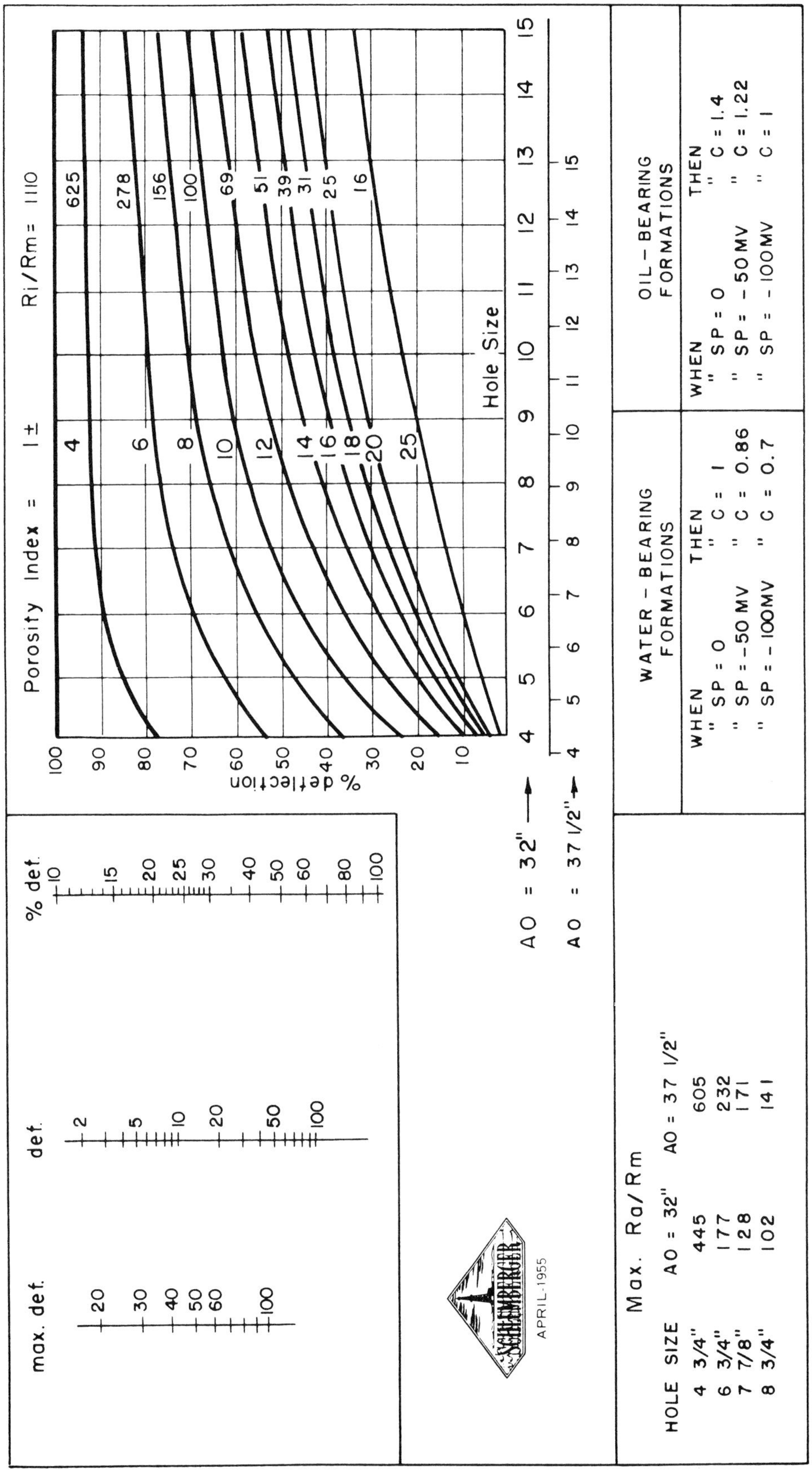
POROSITY INDEX FROM LIMESTONE LATERAL
Porosity Index = I±
Ri/Rm= 1110
% deflection
Hole Size
AO = 32"
AO = 37 1/2"
max. def.
def.
% def.
WATER - BEARING FORMATIONS
WHEN THEN
" SP = 0 " C = 1
" SP = -50 MV " C = 0.86
" SP = -100MV " C = 0.7
OIL - BEARING FORMATIONS
WHEN THEN
" SP = 0 " C = 1.4
" SP = -50MV " C = 1.22
" SP = -100MV " C = 1
Max. Ra/Rm
HOLE SIZE AO = 32" AO = 37 1/2"
4 3/4" 445 605
6 3/4" 177 232
7 7/8" 128 171
8 3/4" 102 141
SCHLUMBERGER
APRIL 1955

Chart C-16

# SCHLUMBERGER WELL SURVEYING CORPORATION

## POROSITY INDEX FROM NEUTRON

*USE AND DESCRIPTION*

Porosity-neutron relations in carbonate rocks have been well established, both through empirical studies and in the laboratory. However, because of the sensitivity of the neutron log to small environmental changes in and around the borehole, the use of laboratory curves for determining porosity has been only partially successful.

This chart presents a technique in which an empirical response line is determined by a maximum deflection minimum porosity point and a pivot point. Using this empirical line, intermediate log readings may be converted to porosity index. Conditions such as hole size, borehole fluid type, mud weight, formation fluid, temperature effects, et cetera, are at least partially normalized.

This method may be used on logs made with the standard $3\frac{5}{8}$" sonde—Types GNAM 1 to 5 equipment, and Types GNT-F or G equipment with 19.5" spacing Ra Be or Pu Be sources. The technique is not valid for logs made with *15.5" spacing GNT-F or G* tools.

Pivot points for GNAM tools in cased and uncased holes are denoted by squares. Pivot points for GNT tools in uncased holes are denoted by circles. Slightly different pivot points are required depending on the type of neutron source used and the rock matrix.

*INFORMATION REQUIRED*

Tool type and neutron source type.
Control point of maximum counting rate for a known minimum porosity.
Neutron deflection in cps or API units.
Hole conditions essentially uniform.
Knowledge of lithology.

*PROCEDURE*

For GNAM equipment: Determine a carbonate control point by finding maximum neutron deflection and assuming that it represents 1% or 2% porosity (based on experience). Draw a straight line from this point through the appropriate GNAM pivot point or through a shale porosity-neutron point. This line gives the neutron-porosity responses for any deflection if hole conditions are unchanged.

For GNT-F or G 19.5" spacing equipment: Determine a control point by finding maximum neutron deflection and assuming that it represents 1% or 2% porosity in carbonates. Draw a straight line from this point through the appropriate GNT-F or G pivot point. This line gives the neutron-porosity response for any deflection if hole conditions remain unchanged.

*EXAMPLE*

GIVEN: GNAM tool
Neutron = 300 cps
Maximum Neutron = 500 cps
(Minimum $\phi$ = 1%)
Uncased hole

SOLUTION: Porosity Index = 10%

GIVEN: GNT-F 19.5" spacing Pu Be source.
Limestone
Neutron = 1500 API units
Maximum Neutron = 3000 API units
(Minimum $\phi$ = 1%)

SOLUTION: Porosity Index = 10%

*NOTE*

Complete laboratory departure curves are available for GNT-F or G 15.5" spacing Ra Be or Pu Be source equipment as well as for GNT-F or G 19.5" spacing Ra Be or Pu Be source equipment.

## POROSITY INDEX FROM NEUTRON

SCHLUMBERGER
OCTOBER 1962

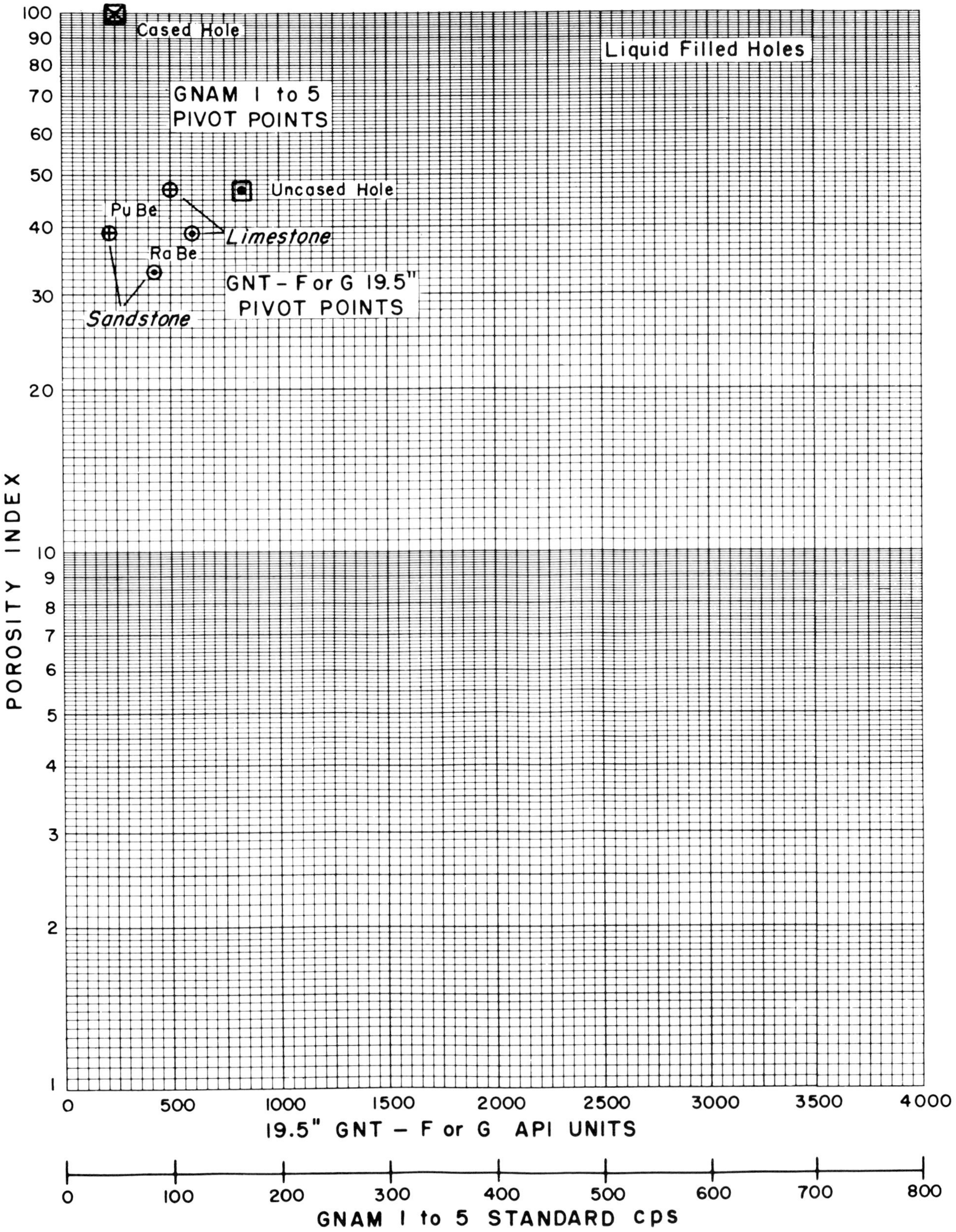


Chart C-18

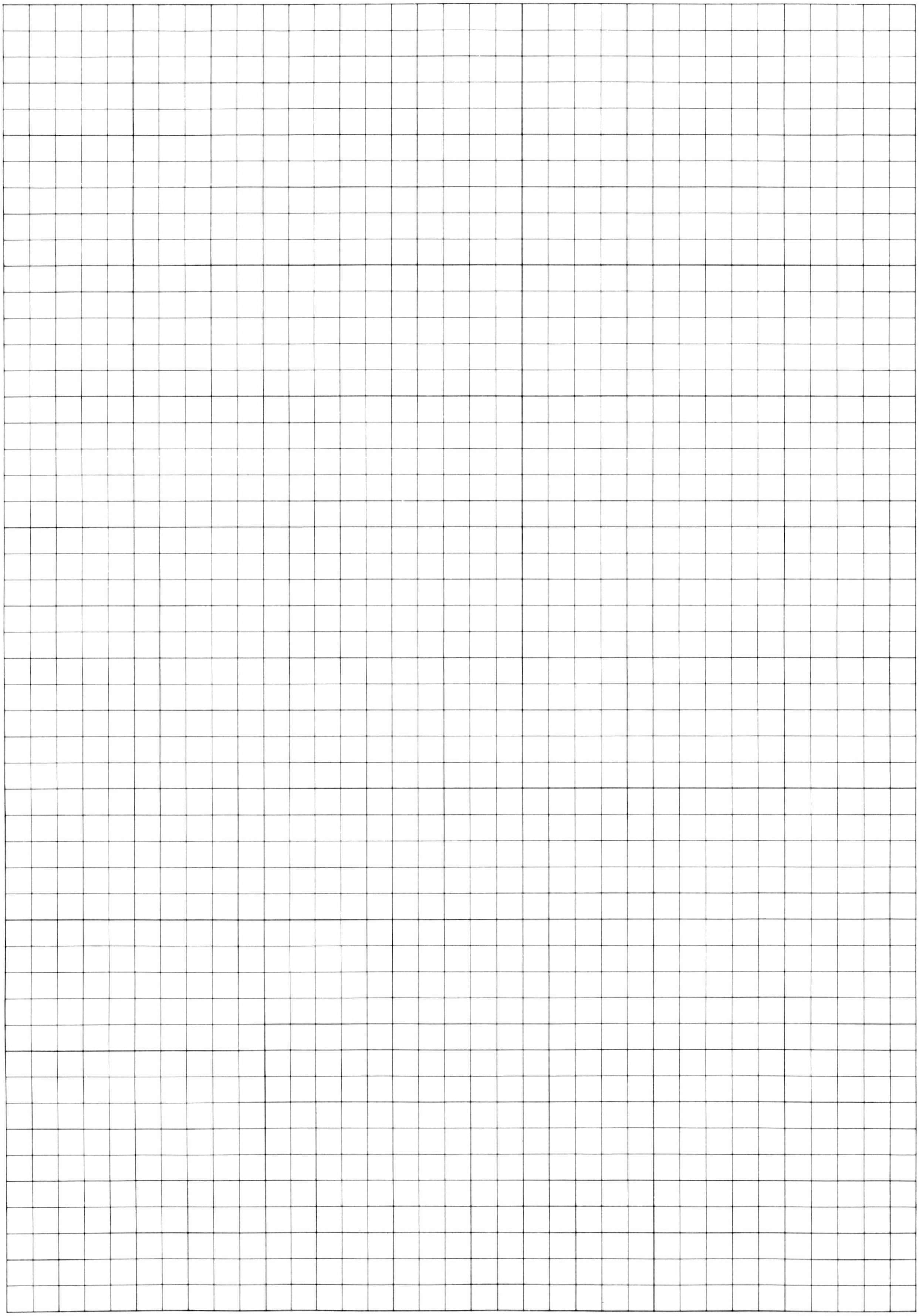

Linear Grid

SCHLUMBERGER

# NEUTRON POROSITY EQUIVALENCE CURVES

## GNT F, G, or H

### (WATER-FILLED HOLES)

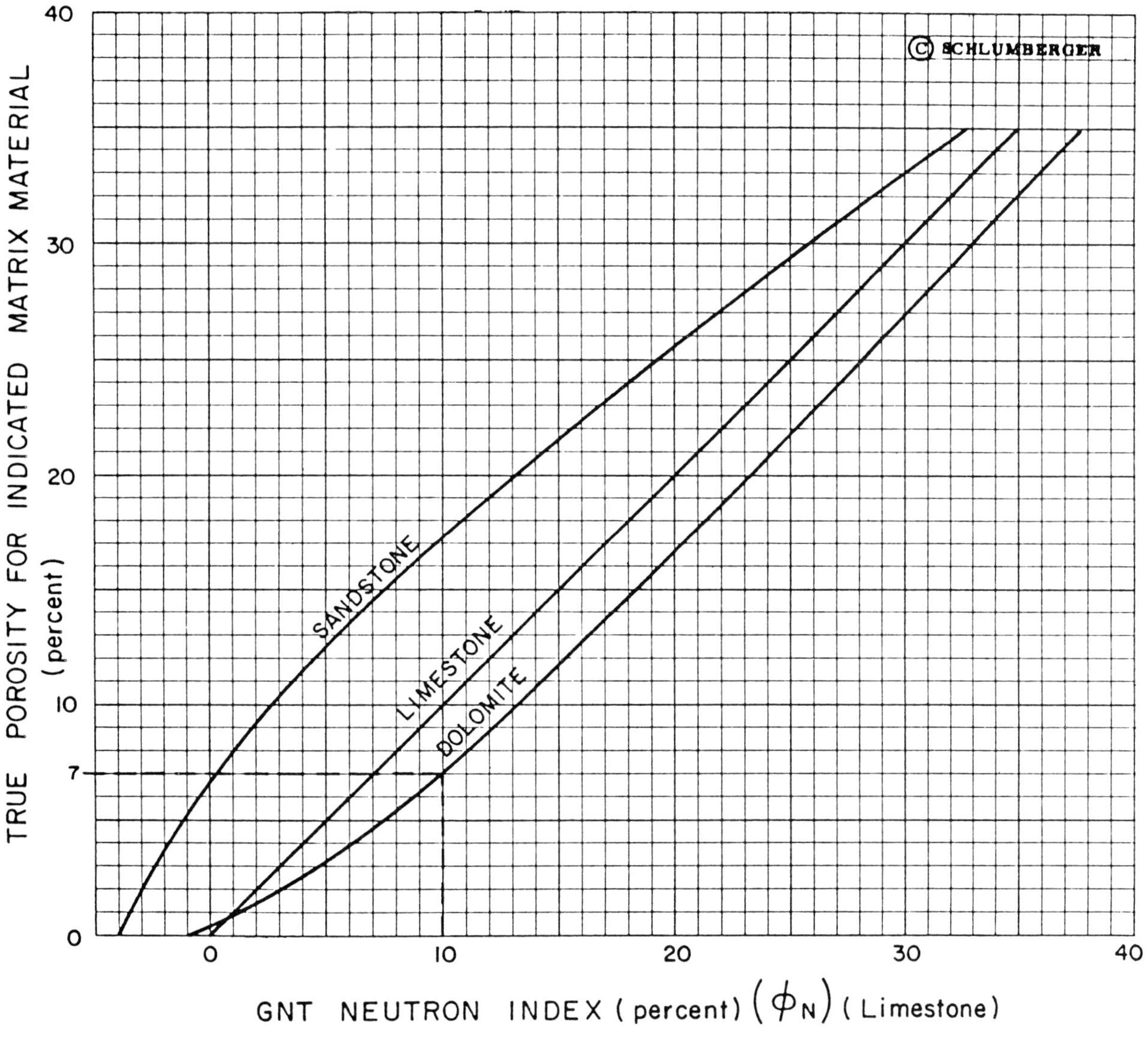


GNT API units are converted to limestone porosity index using the appropriate departure curve. From this value the above chart is used to determine true porosities in sandstones or dolomites.

*Example, Dolomite Bed:* The GNT reads 10 (limestone) porosity units. True porosity equals 7%.

Chart C-19

**SCHLUMBERGER WELL SURVEYING CORPORATION**

# NOMOGRAM FOR SATURATION DETERMINATION—CLEAN FORMATION

## *USE AND DESCRIPTION*

This nomogram provides solutions of the equation $SP = K \log R_{mf}/R_w$ to find $R_w$ and of the Archie relations to find saturation. F or porosity must be known and related through the equation $F = 0.62\phi^{-2.15}$.

## *INFORMATION REQUIRED*

$R_w$ at Formation Temperature

If this is not known, it is found by the use of:

a. Static SP (Deduct electrofiltration if $E_k$ is important.)

b. Formation Temperature or K

c. $R_{mf}$ at Formation Temperature

F—Formation Factor

or

$\phi$—Porosity

$R_t$—True Formation Resistivity

## *PROCEDURE*

*Case Where $R_w$ Is Unknown:* From the static SP and the formation temperature (or K) find $R_{mf}/(R_w)_e$. Then by passing through $R_{mf}$, $(R_w)_e$ is determined. This value may be lower than the true $R_w$ (see pages A-9 through A-12). If greatly different, a true $R_w$ from Chart A-12 should be found before saturation determination. In some regions experience will show that a choice of K, greater than that for the equivalent temperature, will be a more convenient way of obtaining $R_w$ directly, without the need for Chart A-12.

Build $R_o$ by passing on through the known value of F or per cent porosity.
Find per cent $S_w$ by passing on through the known $R_t$.

*Case Where $R_w$ Is Known:* Use of the SP is not required. Enter $R_w$ at formation temperature in Column 4 and proceed as above.

## *EXAMPLE*

GIVEN: $R_t$ = 15.
F = 20.
SP = −100.
$R_{mf}$ = 0.75 at 200°F.

SOLUTION: $S_w$ = 26% ($R_w$ = .05; $R_o$ = 1.0).

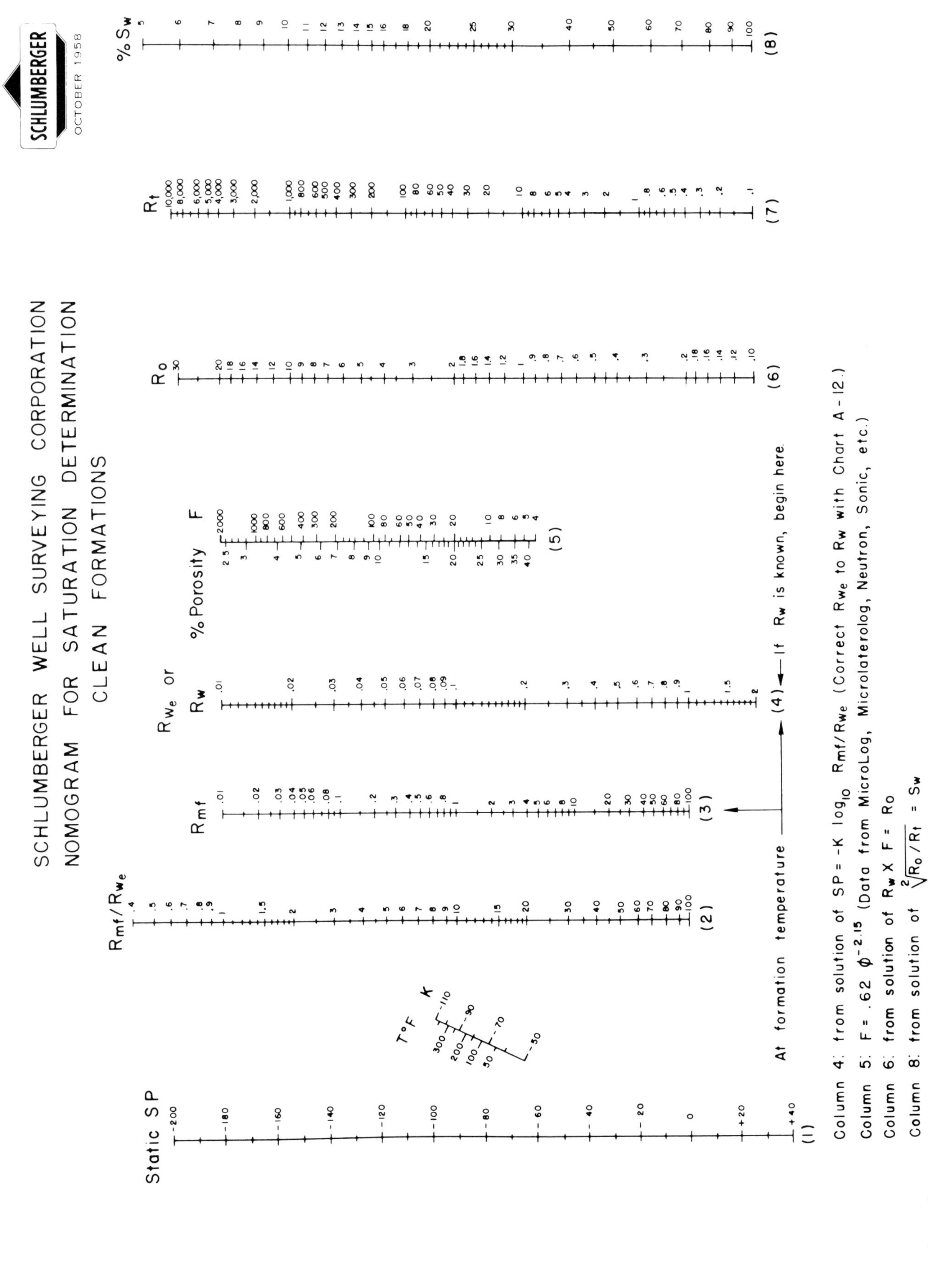
SCHLUMBERGER
OCTOBER 1958
SCHLUMBERGER WELL SURVEYING CORPORATION
NOMOGRAM FOR SATURATION DETERMINATION
CLEAN FORMATIONS
Static SP (1)
T°F K
$R_{mf}/R_{w_e}$ (2)
$R_{mf}$ (3)
$R_{w_e}$ or $R_w$ (4)
% Porosity F (5)
$R_o$ (6)
$R_t$ (7)
% $S_w$ (8)
At formation temperature
If $R_w$ is known, begin here.
Column 4: from solution of SP = -K $\log_{10}$ $R_{mf}/R_{w_e}$ (Correct $R_{w_e}$ to $R_w$ with Chart A-12.)
Column 5: F = .62 $\phi^{-2.15}$ (Data from MicroLog, Microlaterolog, Neutron, Sonic, etc.)
Column 6: from solution of $R_w$ X F = $R_o$
Column 8: from solution of $\sqrt[2]{R_o/R_t}$ = $S_w$

Chart D-2

Linear Grid

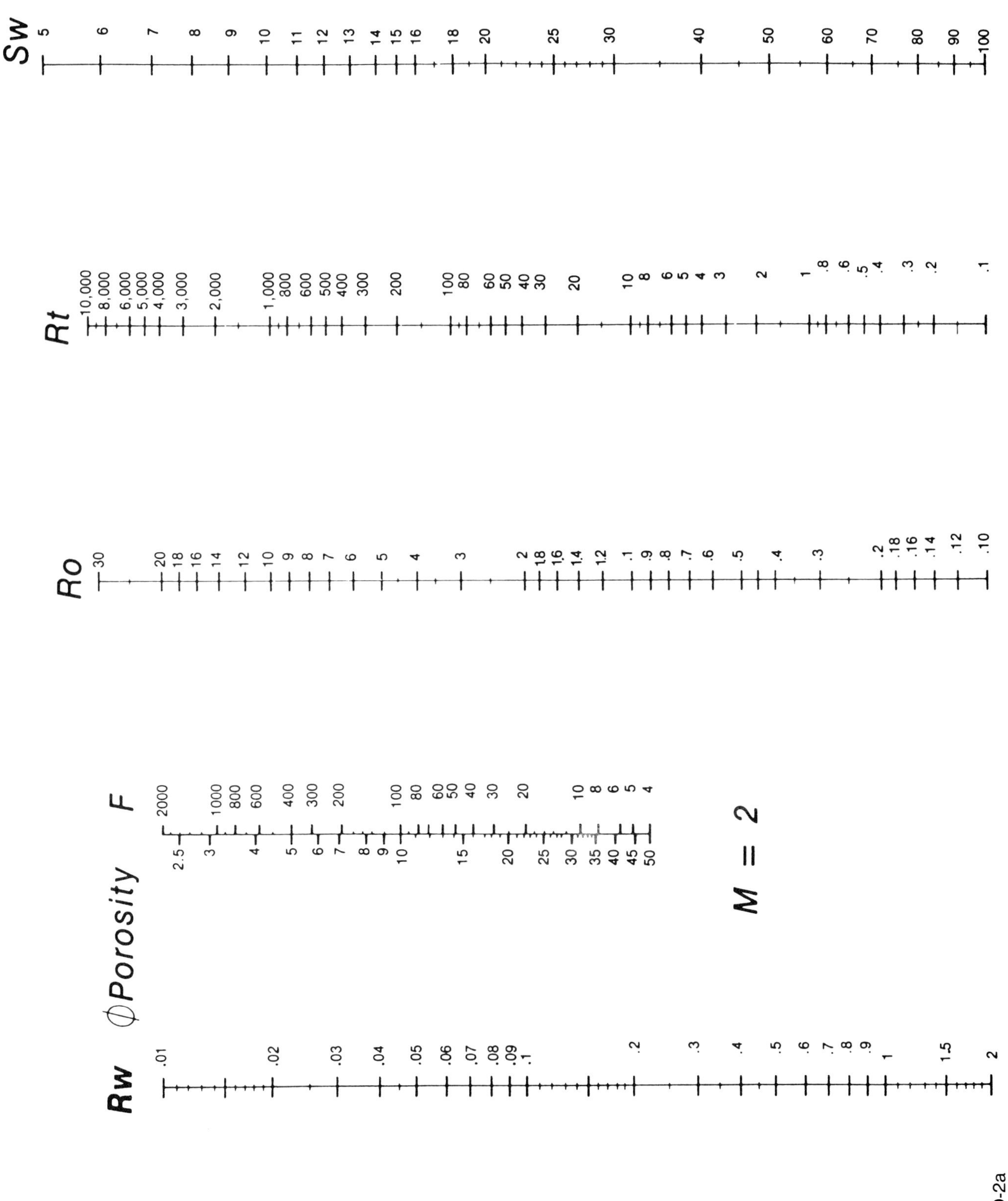


Chart D-2a

SCHLUMBERGER WELL SURVEYING CORP.

# LOG INTERPRETATION CHARTS

## SATURATION DETERMINATION
## CLEAN OR SHALY FORMATION

*USE AND DESCRIPTION*

This chart provides solutions of Equation (6) in the paper by Poupon, Loy, and Tixier, "A Contribution to Electrical Log Interpretation in Shaly Sands", so that $S_w$ in the effective porosity of laminated sands can be found. The error is negligible if the sands also contain disseminated shale. An average Residual Oil (ROS) correction is built in based on the relation:

$$(1 - ROS) = S_{xo} = \sqrt[5]{S_w}.$$

The chart applies equally well to clean sands or limes. In such a case, PSP = SSP; and the apparent $S_w$ is equal to the true $S_w$.

*INFORMATION REQUIRED*

$R_{xo}$ — from the MicroLog
$R_t$ — from the Electrical Log or Induction Log
PSP — read from the SP curve
SSP — read from the SP curve on nearby clean sand
or
$R_{mf}/R_w$ — determined by knowledge of $R_w$ and $R_{mf}$
K — by consideration of formation temperature
ROS — either assumed or known

*PROCEDURE*

Plot the ratio $R_{xo}/R_t$ against PSP. (Make corrections for bed thickness by curve of $R_i/R_m = 5$, Chart A-8, for beds less than 10′ thick.) This point gives an apparent water saturation. Through this point and the circle on the chart where SP = 0 draw a line and extend it until intersection with the appropriate value of SSP or $R_{mf}/R_w$ (scale at top of chart). If you use the ROS assumed, $S_w$ is read by the position between the diagonal lines. If ROS is known, proceed parallel with the diagonal lines to the edge of the chart. Go horizontally to the ROS value and read $S_w$ by the position between the diagonal curved lines.

*EXAMPLE*

I. *Clean Sand*

GIVEN: $R_{xo} = 10.$
$R_t = 5.$
SSP = —100.
K = 80.

SOLUTION: $S_w = 26\%.$
If ROS known to be 20%, $S_w = 28\%.$

II. *Shaly Sand*

GIVEN: $R_{xo}/R_t = 2.5.$
PSP = —65.
SSP = —112.
K = 80.

SOLUTION: Apparent $S_w = 55\%.$
True $S_w = 35\%.$
If SSP not known, $R_{mf}/R_w = 25$ would have been used instead.

NOTE: If extraneous potentials exist, the SP terms have to be reduced accordingly before use in the chart.

Schlumberger

# SATURATION DETERMINATION
## (RATIO METHOD)[12]

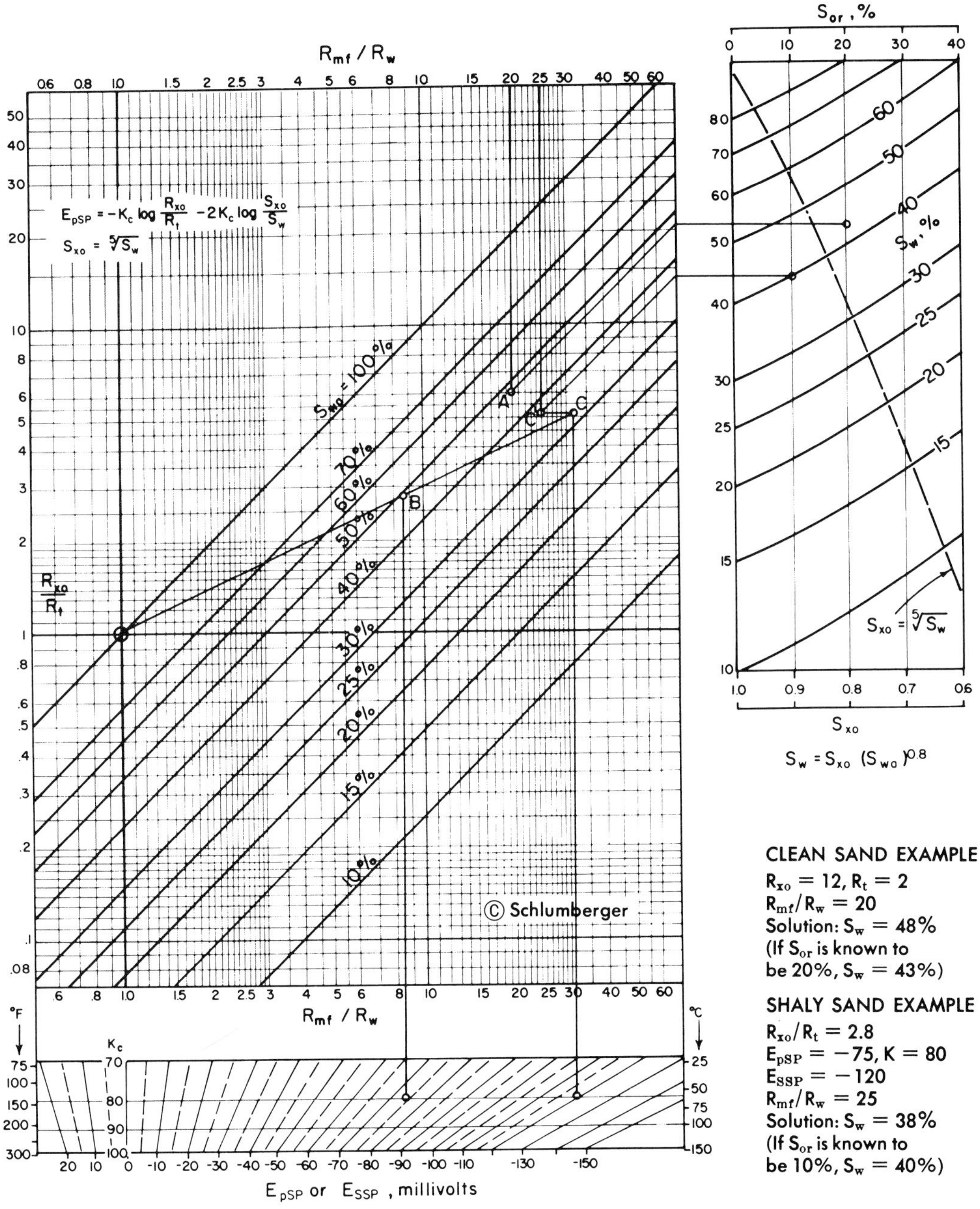


**CLEAN SAND EXAMPLE**

$R_{xo} = 12, R_t = 2$
$R_{mf}/R_w = 20$
Solution: $S_w = 48\%$
(If $S_{or}$ is known to be 20%, $S_w = 43\%$)

**SHALY SAND EXAMPLE**

$R_{xo}/R_t = 2.8$
$E_{pSP} = -75, K = 80$
$E_{SSP} = -120$
$R_{mf}/R_w = 25$
Solution: $S_w = 38\%$
(If $S_{or}$ is known to be 10%, $S_w = 40\%$)

Procedure: For clean sands plot the ratio $R_{xo}/R_t$ against $R_{mf}/R_w$ (Point A) to find water saturation at "average" residual oil saturation ($S_{xo} = S_w^{1/5}$). If $R_{mf}/R_w$ is unknown the SP can be used, assuming $E_{pSP} = -K \log R_{xo}/R_t$; enter chart from SP scale at bottom. (The chart includes no activity correction.)

For shaly sands plot $R_{xo}/R_t$ against $E_{pSP}$ (Point B). Draw a line through this point and the circle where $R_{mf}/R_w = R_{xo}/R_t = 1$. Extend line to intersect the value of $E_{SSP}$ (Point C) to find a value of $R_{xo}/R_t$ approximately corrected for shaliness. Plot this corrected value of $R_{xo}/R_t$ versus $R_{mf}/R_w$ (Point C′) to find $S_w$ assuming "average" $S_{or}$. If $R_{mf}/R_w$ is unknown, use $S_w$ value given by Point C.

Use diagram at right to correct $S_w$ if $S_{or}$ is known.

**Sw-2**

D-4

SCHLUMBERGER WELL SURVEYING CORP.

# LOG INTERPRETATION CHARTS

## POROSITY DETERMINATION — SHALY SAND

*USE AND DESCRIPTION*

This chart provides a method of determining the effective porosity in shaly sands. It is a companion chart to D-4 (Saturation Determination), and it uses much of the same data.

*INFORMATION REQUIRED*

$R_{xo}/R_t$ — which was also used in Chart D-4

$\propto$ — which equals PSP/SSP

$F_a$ — from the ratio $R_{xo}/R_{mf}$

ROS — assumed or estimated

*PROCEDURE*

In the lower part of the chart we find a grid made up of lines for $R_{xo}/R_t$ and $\propto$. $R_{xo}/R_t$ was used in Chart D-4; and we can also use Chart D-4 to find $\propto$, which requires PSP and SSP. If $R_{mf}/R_w$ had been used in place of SSP, the SSP can be found from Chart D-4 by going from $R_{mf}/R_w$ all the way to the bottom, where SSP is read. Then PSP/SSP $= \propto$, which, together with $R_{xo}/R_t$, permits plotting a point within the bottom of this porosity chart.

The point thus established is moved upward to meet a line representing $F_a$. From this position, move horizontally to the right as far as the chosen value of ROS (see scale at top of chart). The location of this point permits interpolation between the diagonal lines of equal effective porosity, so that the desired value, $\phi$, is obtained.

*EXAMPLE*

GIVEN: $R_t = 4.7$; $R_{xo} = 3.75$; PSP = —78; $R_{mf} = .5$; $R_w = .02$; ROS = 10%.

SOLUTION: Find SSP from Chart D-4. Enter $R_{mf}/R_w = 25$ at top; go to bottom of chart and find SSP = —112.

Then, $\propto$ = PSP/SSP = 78/112 = .7.

$R_{xo}/R_t = 3.75/4.7 = .8$.

$F_a = R_{xo}/R_{mf} = 3.75/.5 = 7.5$.

$\phi$, Effective Porosity, = 28%.

(Note that with no correction for shale $\phi$ would have been 35%.)

# POROSITY DETERMINATION
## SHALY SANDS

SCHLUMBERGER
AUGUST-1955

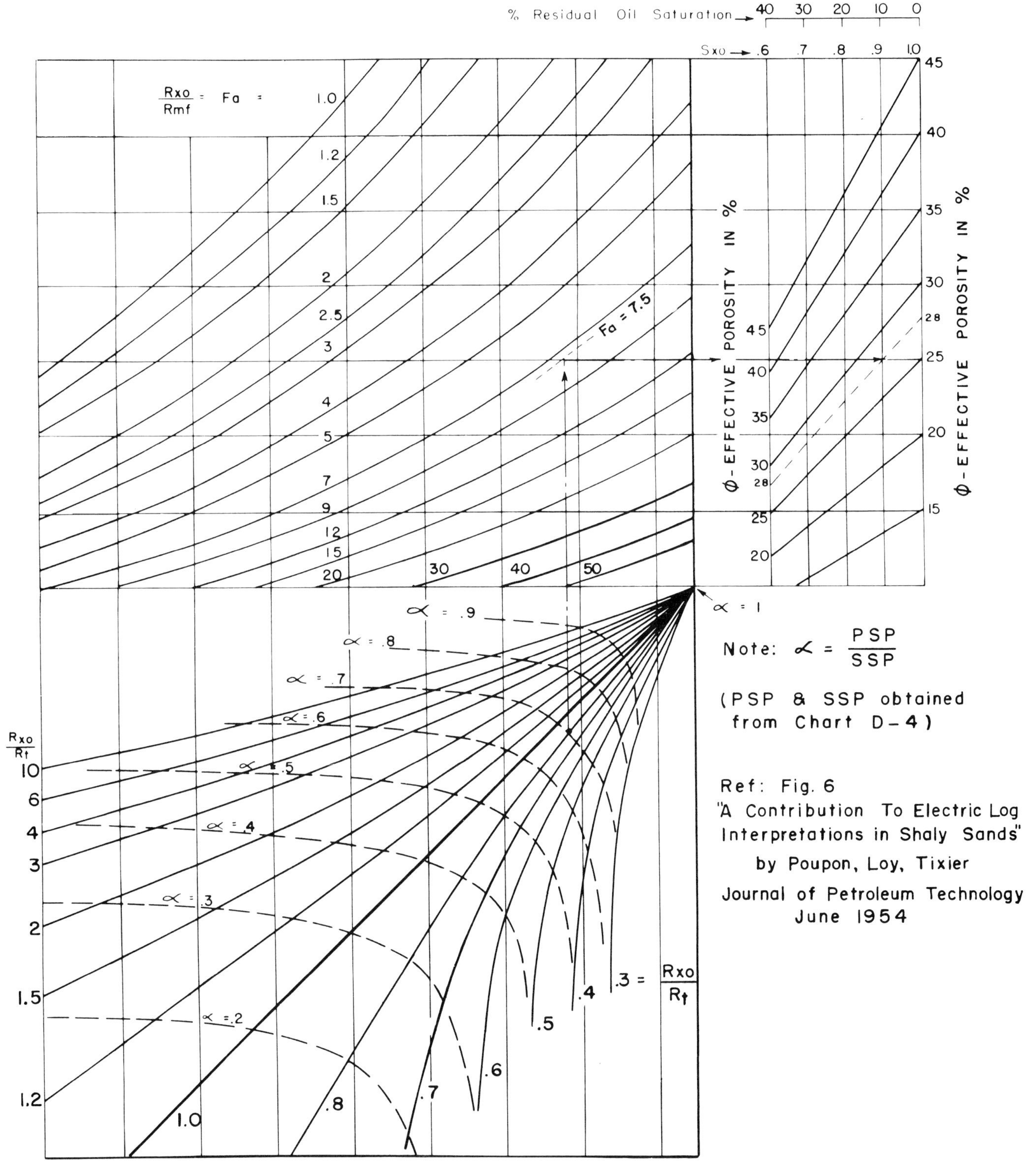


Note: $\alpha = \frac{PSP}{SSP}$

(PSP & SSP obtained from Chart D-4)

Ref: Fig. 6
"A Contribution To Electric Log Interpretations in Shaly Sands"
by Poupon, Loy, Tixier
Journal of Petroleum Technology
June 1954

Chart D-6

SCHLUMBERGER WELL SURVEYING CORP.

# LOG INTERPRETATION CHARTS

## ROCKY MOUNTAIN METHOD

### *USE AND DESCRIPTION*

This empirical method permits saturation and porosity determination without prior knowledge of the formation factor or $R_0$. It is essential that a definite invaded zone be present; therefore, an essential requirement for its application is that $R_{16''}/R_m$ be at least 10.

The chart is of two parts:

1. *Saturation* is found by solution of the equation: $S_w = \frac{R_i/R_t}{R_z/R_w}$.

   The solution is realized by plotting $R_i/R_t$ versus the SP, which controls the value of $R_z/R_w$.

2. *Porosity* is found by solution of the Archie Formula in the form:

   $F = S_w^2 \frac{R_t}{R_w}$, and F is then related to porosity through the Humble Formula.

### *INFORMATION REQUIRED*

$R_m$ at Formation Temperature
d — hole size
$R_{16''}$ and $R_{18'8''}$ corrected by Chart B-2 to give $R_i/R_m$ and $R_t/R_m$
SP and an approximate value of K

### *PROCEDURE*

1. $R_i/R_m$ is divided by $R_t/R_m$ to get the ratio $R_i/R_t$. This value is entered in the upper part of the chart on the diagonal system of lines referred to the SP thus locating a point. From there, go vertically downward to the saturation scale to get $S_w$.

2. From the SP, the value of $R_m/R_w$ is estimated by the scale on the upper left. $R_m/R_w$ times $R_t/R_m$ gives the required $R_t/R_w$, which is entered on the lower left scale. The $S_w$ value is carried down to meet $R_t/R_w$, and the intersection enables porosity estimation from reference to the diagonal system of lines.

### *EXAMPLE*

GIVEN: $R_m = .5$ at Formation Temperature (190°F).
$d = 9''$.
$SP = —40$ MV.
$R_{16''} = 50$.
$R_{18'8''} = 250$.
From Chart B-2 we found $R_i/R_m = 150$ and $R_t/R_m = 300$.

SOLUTION:

1. ***Saturation***
   $R_i/R_t = R_i/R_m \div R_t/R_m = 150 \div 300 = 0.5$.
   $SP = —40$ with K assumed as 90.
   $S_w = 20\%$.

2. ***Porosity***
   From the SP (—40 MV, $K = 90$), we can see that
   $R_m/R_w = 3$.
   Then, $R_t/R_w = R_t/R_m \times R_m/R_w = 3 \times 300 = 900$.
   Porosity = 15%.

NOTE:

It should be emphasized that caution must be taken when $R_i/R_t$ is close to unity since it may then be possible that the formation is very tight and extremely deeply invaded, or that it is very dense and impermeable. On the other hand, many good reservoirs will show $R_i/R_t \simeq 1.0$, due to resistivity compensation during the invasion of oil sands.

For discussion of the details concerning this method and its applications, the reference cited on the chart should be read.

SCHLUMBERGER

OCTOBER 1962

# ROCKY MOUNTAIN METHOD FOR $S_w$

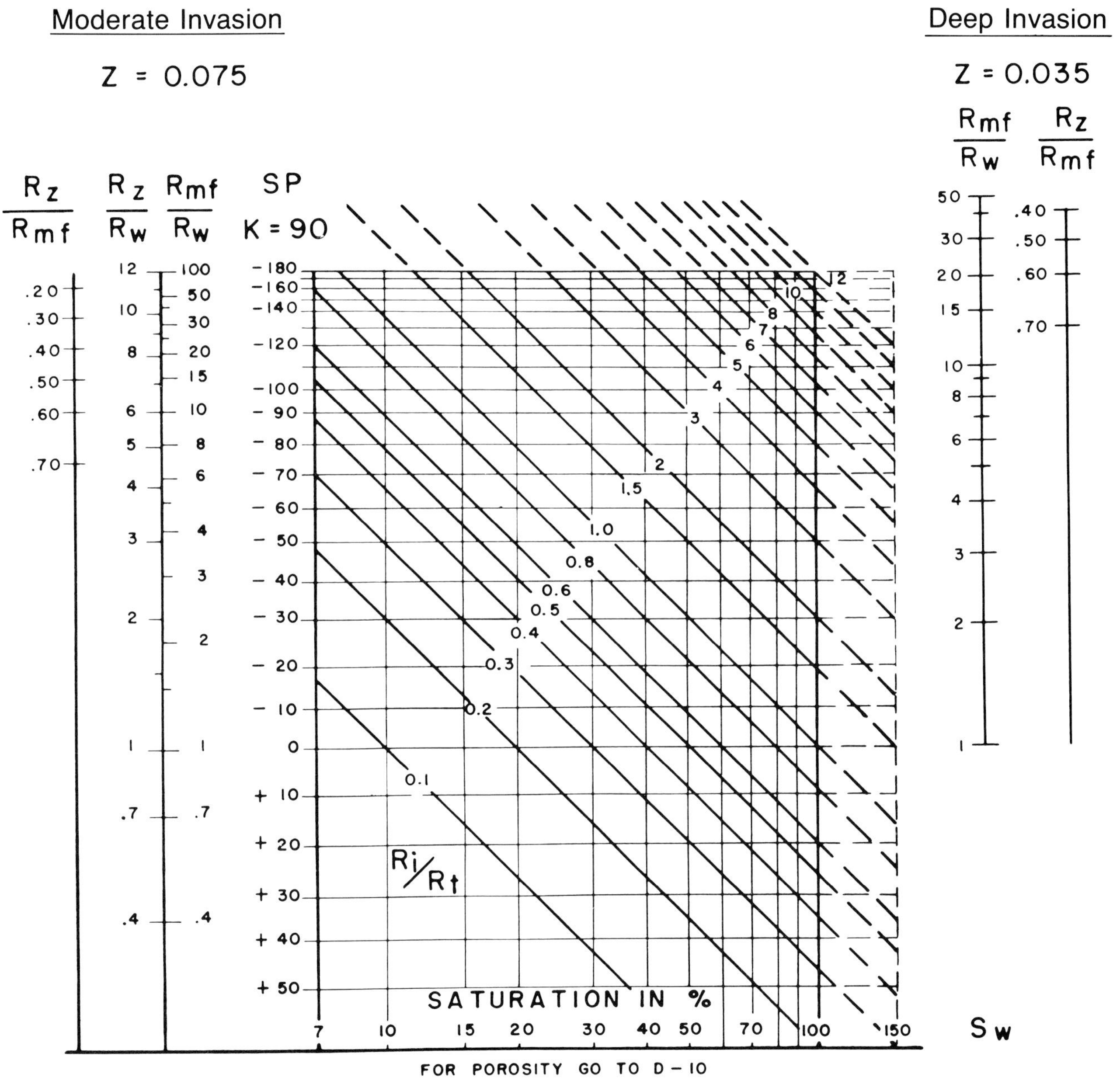


REFERENCE: Tixier, M. P. "Electric Log Analysis in the Rocky Mountains." *Oil & Gas Journal*, June 23, 1949.

Chart D-8

SCHLUMBERGER WELL SURVEYING CORP.

## POROSITY BALANCE

*USE AND DESCRIPTION*

The purpose of porosity balance is to verify water saturations derived from resistivity ratio methods (Charts D-8, D-12, and D-14). Inferences can also be made concerning invasion and the accuracy of certain log values.

The balance consists of solving the Archie saturation formula in reverse to see if the porosity (or F) so derived matches information from an independent source. Confirmation indicates the $S_w$ is correct. If the F computed by the porosity balance is too large, the $S_w$ is also too large and vice versa. (See reference below for details.)

*INFORMATION REQUIRED*

$R_w$ and $R_{mf}/R_w$; $R_z$ should be found by use of the alignment chart on D-8
$R_L$ — long spacing, which is assumed to be $R_t$
$R_S$ — short spacing, which is assumed to be $R_i$ (or $R_{16c}$)
$Ø_t$ — porosity from an independent method
$S_{wc}$ — apparent $S_w$ from a resistivity ratio chart

*PROCEDURES*

**When Rocky Mountain Chart (D-8) Used**

A) If $Ø_c > Ø_t$: $S_{wc} < S_w$

If an Induction was used for $R_L$, when $R_{16c}/R_{5FF40} < 1.2$ or $R_{16c}/R_{6FF40} < 1.4$, use $R_L$ as $R_t$ in Archie method. If the ratios are greater, and for ES logs, the error may be found as follows:

1) If $FR_z > R_s$, invasion is small, and $R_t$ is verified; use the Archie method to find $S_w$. Move horizontally to correct $Ø_t$.

2) If $FR_z \leq R_s$, invasion is deep, $R_L$ is higher than $R_t$. Move the point located by $S_{wc}$ and $R_t/R_w$ at 45° until $Ø_t$ is met; read new value of $S_w$. In cases of high SP, this $S_w$ should be less than $R_z/R_{mf}$ to verify that oil is present. The scale below is used to find $R_z/R_{mf}$.

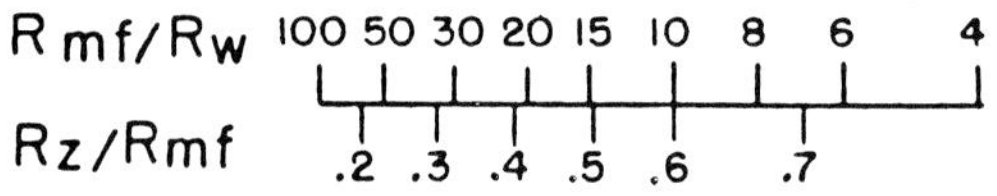


B) If $Ø_c < Ø_t$: $S_{wc} > S_w$.

This occurs when (1) the $R_L$ is too low; e.g., taken from the Induction with conditions of annulus or $R_{xo} < R_t$, or (2) when $R_S$ is too high due to adjacent high resistivity.

For Case 1 use the $R_i$ method (45° displacement).

For Case 2 use the Archie method using $R_L$ as $R_t$. (Horizontal displacement)

**When IES Chart (D-12, D-14) Used**

A) If $Ø_c > Ø_t$: $S_{wc} < S_w$

If an Induction was used for $R_L$, when $R_{16c}/R_{5FF40} < 1.2$ or $R_{16c}/R_{6FF40} < 1.4$, use $R_L$ as $R_t$ in Archie method. If the ratios are greater, the error may be found as follows:

1) If $FR_z > R_s$, invasion is small, and $R_L$ is verified as $R_t$. Use Archie method $R_{IL}$ as $R_t$.

2) If $FR_z \leq R_s$, invasion is deep and $R_{IL}$ is too high. On the IES chart (D-12 or D-14) move vertically to the critical saturation line, finding a new $R_L$, then enter the porosity balance chart and check porosities, using $S_w = 60\%$ (6FF40) or 50% (5FF40). If $Ø_c < Ø_t$, you have overcorrected and must iterate to find $R_t$ and thence $S_w$. If $Ø_c > Ø_t$, you know that $S_w > 60\%$, and the zone is probably water.

B) If $Ø_c < Ø_t$: $S_{wc} > S_w$.

If shoulder beds have not made $R_s$ too high, you may have a case of annulus or $R_{xo} < R_t$. If so, take $R_s$ as $R_t$ and find $S_w$ by the Archie method.

*EXAMPLE*

GIVEN:

$R_{IL} = 10 = R_L$
$R_{16c} = 20 = R_S$
$R_{mf}/R_w = .8/.02 = 40$
$Ø_t = 14$ ($F = 42$)
$S_{wc} = 19\%$ (Chart D-8)

SOLUTION:

$R_t/R_w = 10/.02 = 500$
$Ø_c = 21\%$ (does not balance with $Ø_t = 14\%$)
$R_z = 10.3\ R_w$ (from Chart D-8) $= .21$
$FR_z = 42 \times .21 = 8.8$; since
$R_{16c} > 8.8$, invasion is not shallow,
and $R_t$ probably too high.

Moving the plotted point at 45°, we get $S_w = 46\%$; but $R_Z/R_{mf} = .25$, so this zone is probably wet. ($S_w < R_Z/R_{mf}$ to confirm pay.)

*SAME EXAMPLE,* USING CHART D-12

GIVEN:

$R_{16c}/R_{IL} = 2.0$ and since
$R_{mf}/R_w = 40$, $S_{wc} \simeq 20\%$
$Ø_t = 14\%$ ($F = 42$)

SOLUTION:

$R_t/R_w$ and $S_{wc}$ are entered in the Ø balance chart, and $Ø_c = 20\%$. The $FR_Z$ check shows invasion is not shallow so $R_{IL}$ was wrong (too high).

For $R_{mf}/R_w = 40$, the critical line on D-12 gives $R_{16c}/R_{IL} = 8$, thus $R_{IL}$ is computed to be $R_{16c}/8 = 2.5$. The $S_{wc}$ is now 50%.

Entering into the Ø balance chart $S_{wc} = 50\%$, and $R_t/R_w = 2.5/.02 = 125$ gives $Ø_c = 17\%$. This is still too high, so $S_w > 50\%$ and the zone is water.

REFERENCE: Tixier, M. P. "Porosity Balance Verifies Water Saturation Determined from Logs." *Journal of Petroleum Technology,* July, 1958.

# POROSITY DETERMINATION

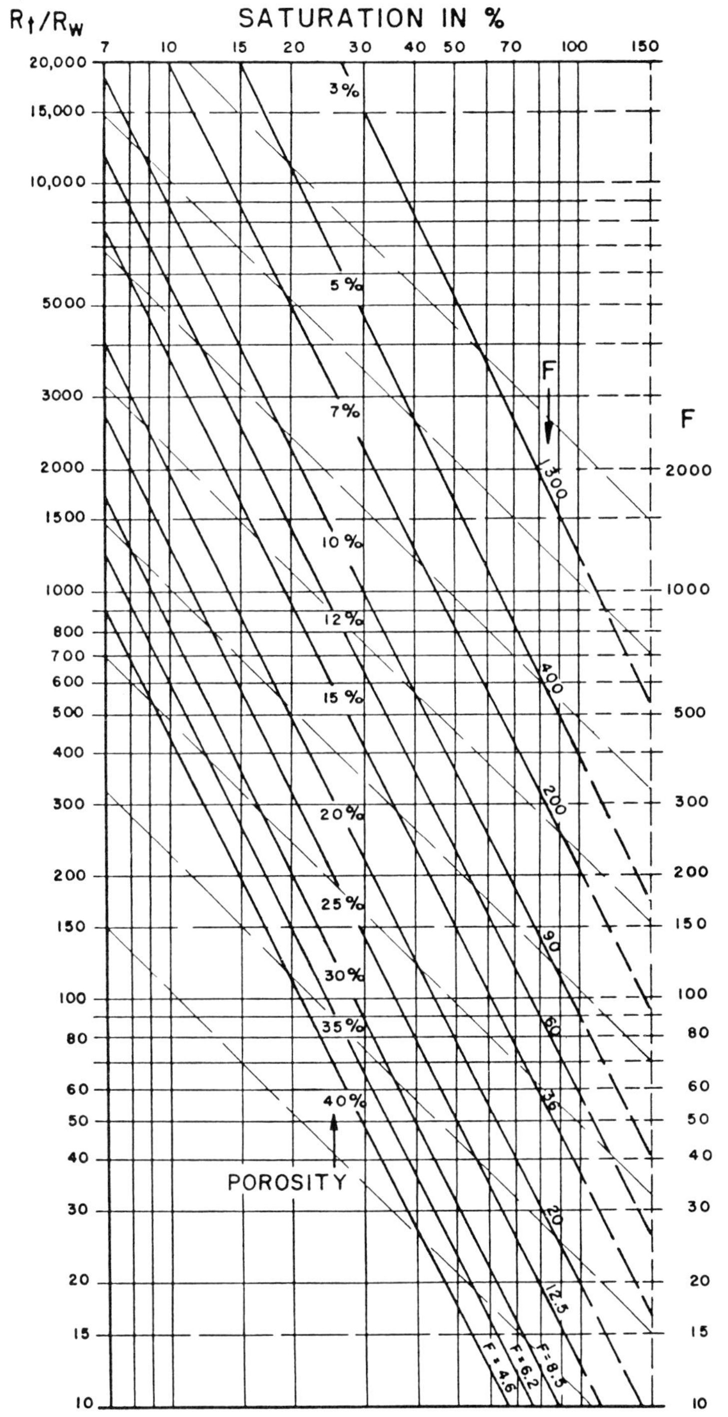


August, 1960

## POROSITY BALANCE FOR ROCKY MOUNTAIN METHOD

If balance does not exist, move from the calculated porosity.

a) along a 45° line until the correct porosity is met (case of $R_t$ wrong),* ($R_i$ Method)

b) along a horizontal line until the correct porosity is met (case of $R_i$ wrong), i. e. $FR_z > R_{16''}$ (Archie Method)

This final point indicates a more correct $S_w$.

<u>Beware of high SP</u>

<u>when the final $S_w \geq$</u> $\frac{R_z}{R_{mf}}$ .

<u>The zone may be wet.</u>

*Exception: When Induction is used, and $R_{16c}/R_{IL} < 1.2$ for 5FF40, and $< 1.4$ for 6FF40, $R_{IL} \simeq R_t$.

Chart D-10

# SCHLUMBERGER WELL SURVEYING CORPORATION

## LATEROLOG INTERPRETATION–$S_w$

*USE AND DESCRIPTION*

This chart provides a ready means of computing $S_w$ when the Laterolog is used for measurement of $R_t$. Incorporated is a means of correcting the Laterolog value for the effects of invasion.

As constructed the correction of the Laterolog reading for invasion is based upon ROS = 20% in the flushed zone.

*INFORMATION REQUIRED*

$\phi$ – determined from Sonic or Neutron Log
$m$ – cementation exponent
} or: F – determined from Microlaterolog.

$R_{mf}$ – at formation temperature

$R_{LL}$ – Resistivity value from Laterolog

Type Laterolog device – indicated on heading

$D_i$ – assumed invasion diameter*

$R_{mf}/R_w$

*PROCEDURE*

*If porosity is obtained from Sonic or Neutron Logs:*

From the porosity value on the left of the chart extend a straight line through the appropriate value of $m$ to the line for F; from there draw a line through the value of $R_{mf}$ at formation temperature to the line $FR_{mf}$. From this intercept extend a line through $R_{LL}$ to the line $\frac{FR_{mf}}{R_{LL}}$. From this point parallel the nearest diagonal line to the assumed invasion diameter for the appropriate Laterolog device. From the intercept on the assumed diameter of invasion proceed horizontally to the line for $\frac{FR_{mf}}{R_t}$. Draw a line from this point through the appropriate $R_{mf}/R_w$ value to the solution of $S_w$.

*If F is obtained from Microlaterolog:*

Enter the chart directly on the line for F and proceed to the solution as described above.

*EXAMPLE*

GIVEN: Porosity from Sonic Log = 13%
$m$ = 2.0
$R_{mf}$ = .08
$R_{LL}$ = 18. (Laterolog 7)
$D_i$ = 5d
$R_{mf}/R_w$ = 1.2

SOLUTION: $S_w$ = 41%.

*For a discussion of invasion effects on the Laterolog, see Williston Basin Symposium paper, "New Logging Approaches for Evaluation of Carbonate Reservoirs," by B. M. McVicar, J. L. Heath, and R. P. Alger, 1956.

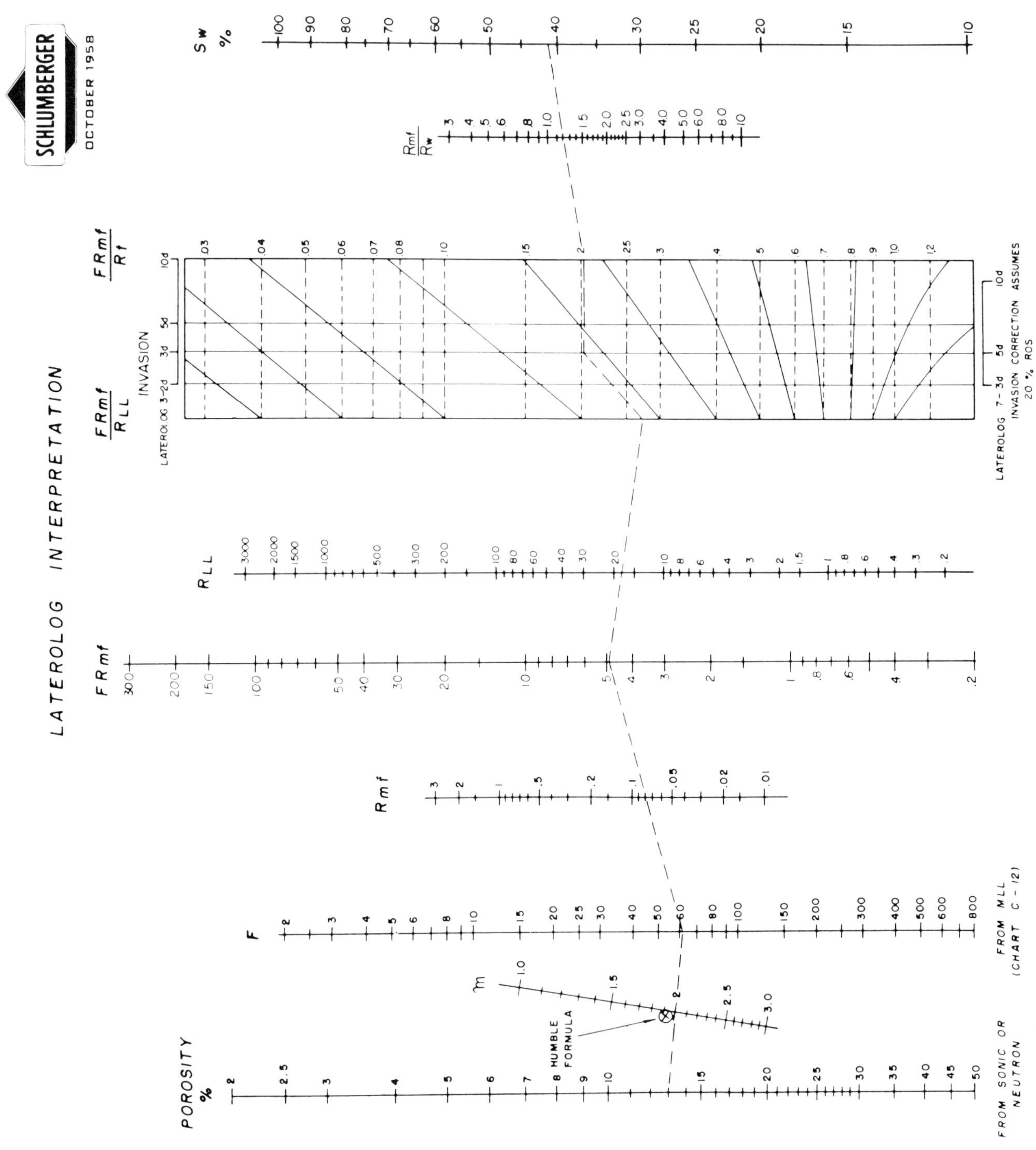
SCHLUMBERGER
OCTOBER 1958
LATEROLOG INTERPRETATION
POROSITY %
FROM SONIC OR NEUTRON
m
HUMBLE FORMULA
F
FROM MLL (CHART C - 12)
Rmf
FRmf
RLL
FRmf / RLL
INVASION
LATEROLOG 3-2d
FRmf / Rt
LATEROLOG 7 - 3d
INVASION CORRECTION ASSUMES 20 % ROS
Rmf / Rw
Sw %

Chart D-16

# SCHLUMBERGER WELL SURVEYING CORPORATION

## WATER SATURATION FROM THE THERMAL DECAY TIME LOG

*USE AND DESCRIPTION*

This nomograph provides a means of determining $S_w$ in cased holes. Optimum interpretation results are obtained in medium to high porosity formations that are neither invaded nor excessively shaly, and when formation water salinity exceeds 80,000 ppm.

*INFORMATION REQUIRED*

*In Clean Formations*

$\tau_{log}$ – thermal decay time, read from log.

$\tau_m$ – formation matrix thermal decay time. For sandstones, varies from 325 to 650 μsecs, and is commonly found to be 500 μsecs. For carbonates, varies from 480 to 610 μsecs, and is commonly found to be 550 μsecs.

$\phi$ – formation porosity

Water Salinity – in ppm

Type of Hydrocarbon – oil or gas. If gas, the approximate pressure is also required

*In Shaly Formations*

All of the above, plus:

p – shale percentage of formation bulk volume.

$\tau_{sh}$ – shale thermal decay time. If not known, this value may be read from the log in an almost-pure-shale section.

*PROCEDURE*

*For clean formations: First,* construct a "matrix" line from pivot point "X" to the value of $\tau_m$ on Column 2. *Second,* establish the point that is the intersection of the "matrix" line and appropriate "porosity" line. *Third,* enter the value from the log, $\tau_{log}$, on Column 2, and, from this point, draw a line through the "matrix-porosity" intersection to Column 3. *Fourth,* establish the point between Columns 3 and 4 described by the "Formation Water Salinity" and the appropriate hydrocarbon line. *Fifth,* construct a line through this point from the intersection on Column 3, and extend this line to Column 4. *Sixth,* read the value of $S_w$ at this intersection on Column 4.

*For shaly formations:* The procedure for shaly sands is the same as for clean formations, except that $\tau_{log}$ must first be corrected for shale effect. This correction is performed as follows: *First,* construct a "shale line" from pivot point "Y" to the value of $\tau_{sh}$ on Column 1. *Second,* establish the point that is the intersection of the "shale line" and the appropriate shale percentage "p". *Third,* enter the value from the log, $\tau_{log}$, on Column 1 and, from this point, construct a line through the "shale line"–"percentage shale" intersection to Column 2. This intersection with Column 2 is the value of $\tau_{log}$, corrected for shale effect, to be used in the determination of $S_w$. The procedure from this point on is the same as previously described for clean formations.

*EXAMPLES*

*Clean Formations*

| GIVEN: | | |
|---|---|---|
| $\tau_{log}$ | = | 300 μsec. |
| $\tau_m$ | = | 500 μsec. |
| $\phi$ | = | 30% |
| Hydrocarbon | = | oil |
| Salinity | = | 100,000 ppm |

SOLUTION: $S_w$ = 22%

*Shaly Formation*

| GIVEN: | | |
|---|---|---|
| $\tau_{log}$ | = | 250 μsec. |
| $\tau_{sh}$ | = | 150 μsec. |
| p | = | 20% |
| $\tau_m$ | = | 500 μsec. |
| $\phi$ | = | 30% |
| Hydrocarbon | = | oil |
| Salinity | = | 100,000 ppm |

SOLUTION: $S_w$ = 22%

NOTE: For those who are used to working with neutron capture cross section, Σ, rather than with τ, a convenient bar nomogram relating these parameters is presented below. It is based on the relation $\tau = 4.55/\Sigma$.

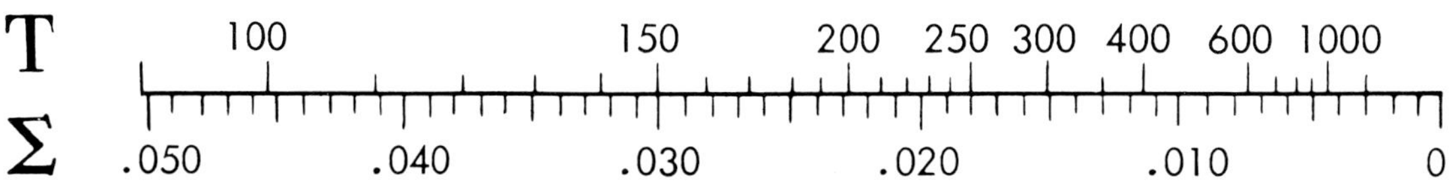

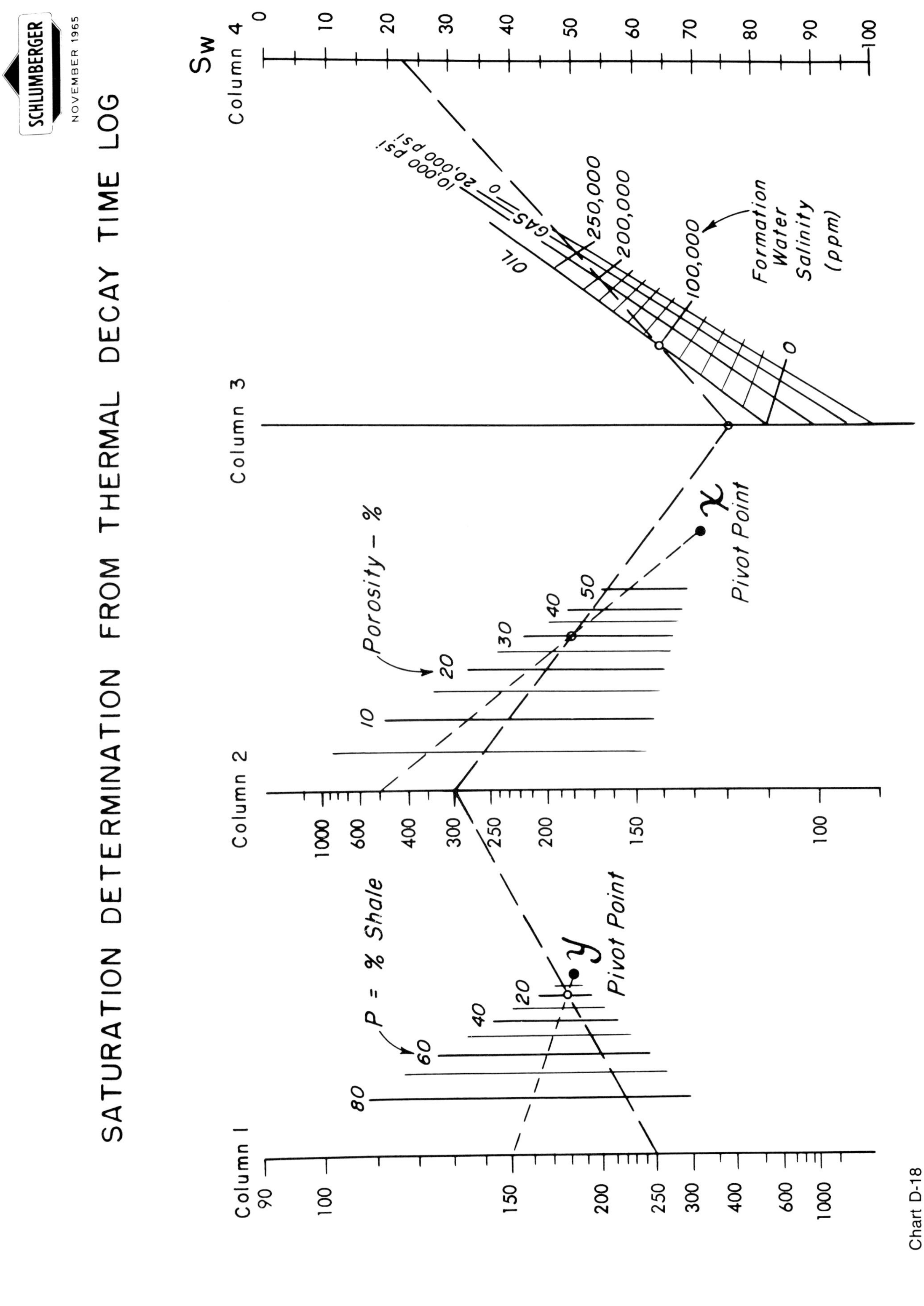
SCHLUMBERGER
NOVEMBER 1965
SATURATION DETERMINATION FROM THERMAL DECAY TIME LOG
Column 1
90
100
150
200
250
300
400
600
1000
P = % Shale
80
60
40
20
y Pivot Point
Column 2
1000
600
400
300
250
200
150
100
Porosity – %
10
20
30
40
50
x Pivot Point
Column 3
OIL
GAS
0
10,000 psi
20,000 psi
250,000
200,000
100,000
0
Formation Water Salinity (ppm)
Column 4
Sw
0
10
20
30
40
50
60
70
80
90
100

Chart D-18

# B

# Formation Tester Interpretation Charts

## CONTENTS

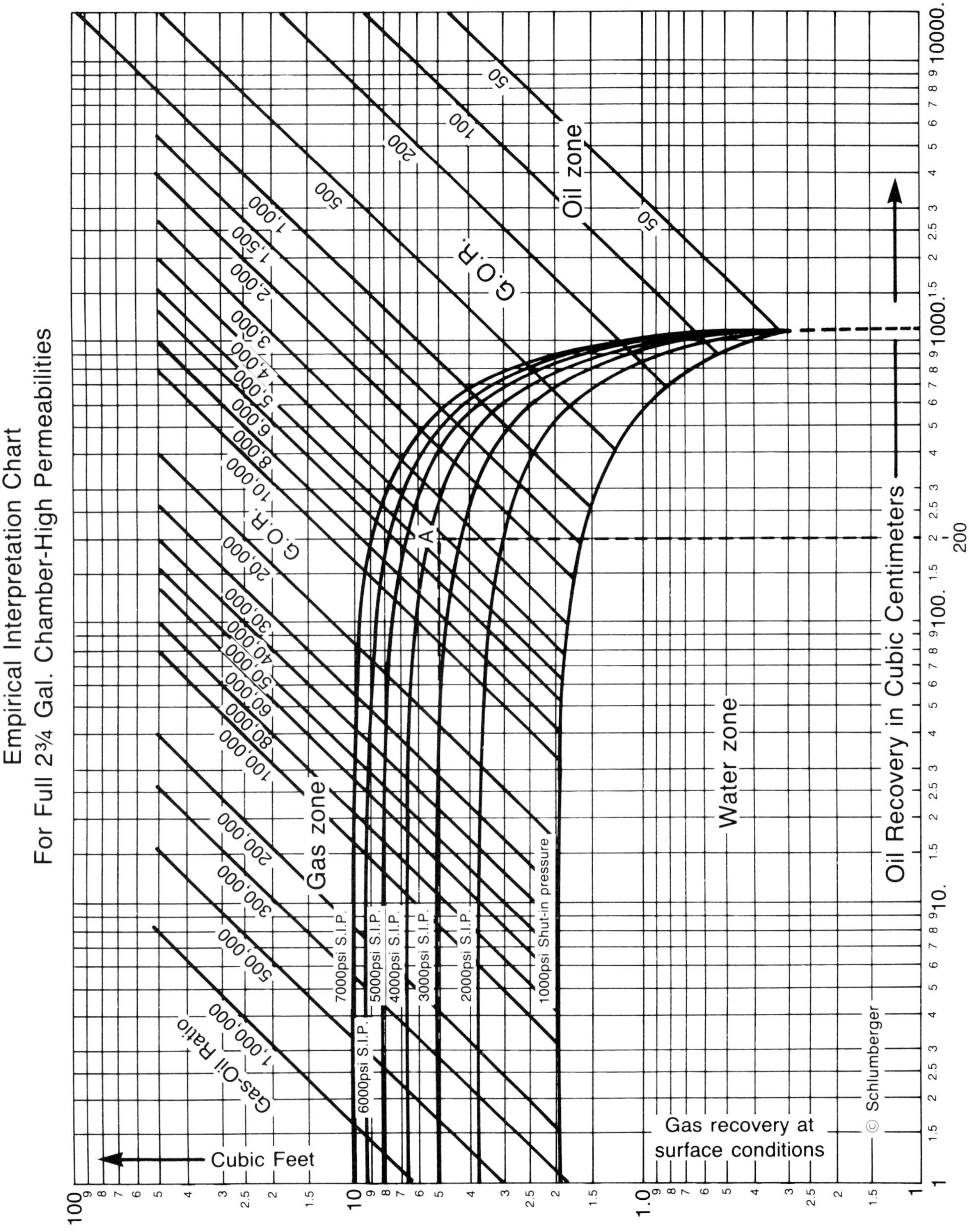
Empirical Interpretation Chart
For Full 2¾ Gal. Chamber-High Permeabilities
Gas-Oil Ratio
1,000,000
500,000
300,000
200,000
100,000
80,000
60,000
50,000
40,000
30,000
20,000
G.O.R.
10,000
8,000
6,000
5,000
4,000
3,000
2,000
1,500
1,000
500
200
100
50
G.O.R.
Oil zone
50
Gas zone
Water zone
A
7000psi S.I.P.
6000psi S.I.P.
5000psi S.I.P.
4000psi S.I.P.
3000psi S.I.P.
2000psi S.I.P.
1000psi Shut-in pressure
Cubic Feet
Gas recovery at surface conditions
© Schlumberger
Oil Recovery in Cubic Centimeters
200

Chart FT-1

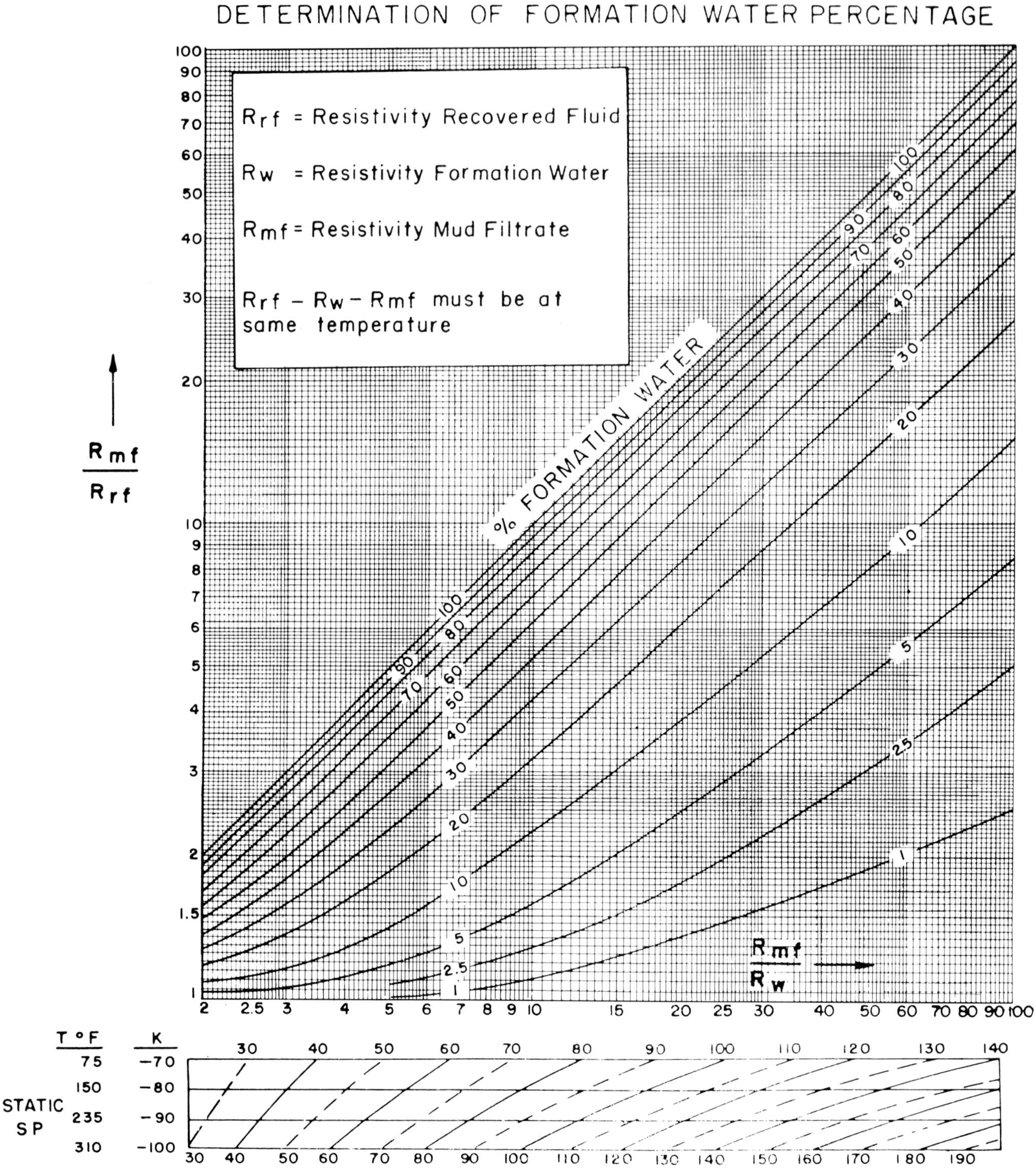
DETERMINATION OF FORMATION WATER PERCENTAGE
Rrf = Resistivity Recovered Fluid
Rw = Resistivity Formation Water
Rmf = Resistivity Mud Filtrate
Rrf – Rw – Rmf must be at same temperature
% FORMATION WATER
Rmf / Rrf
Rmf / Rw
STATIC SP
T °F
K
75 –70
150 –80
235 –90
310 –100

Chart FT-2

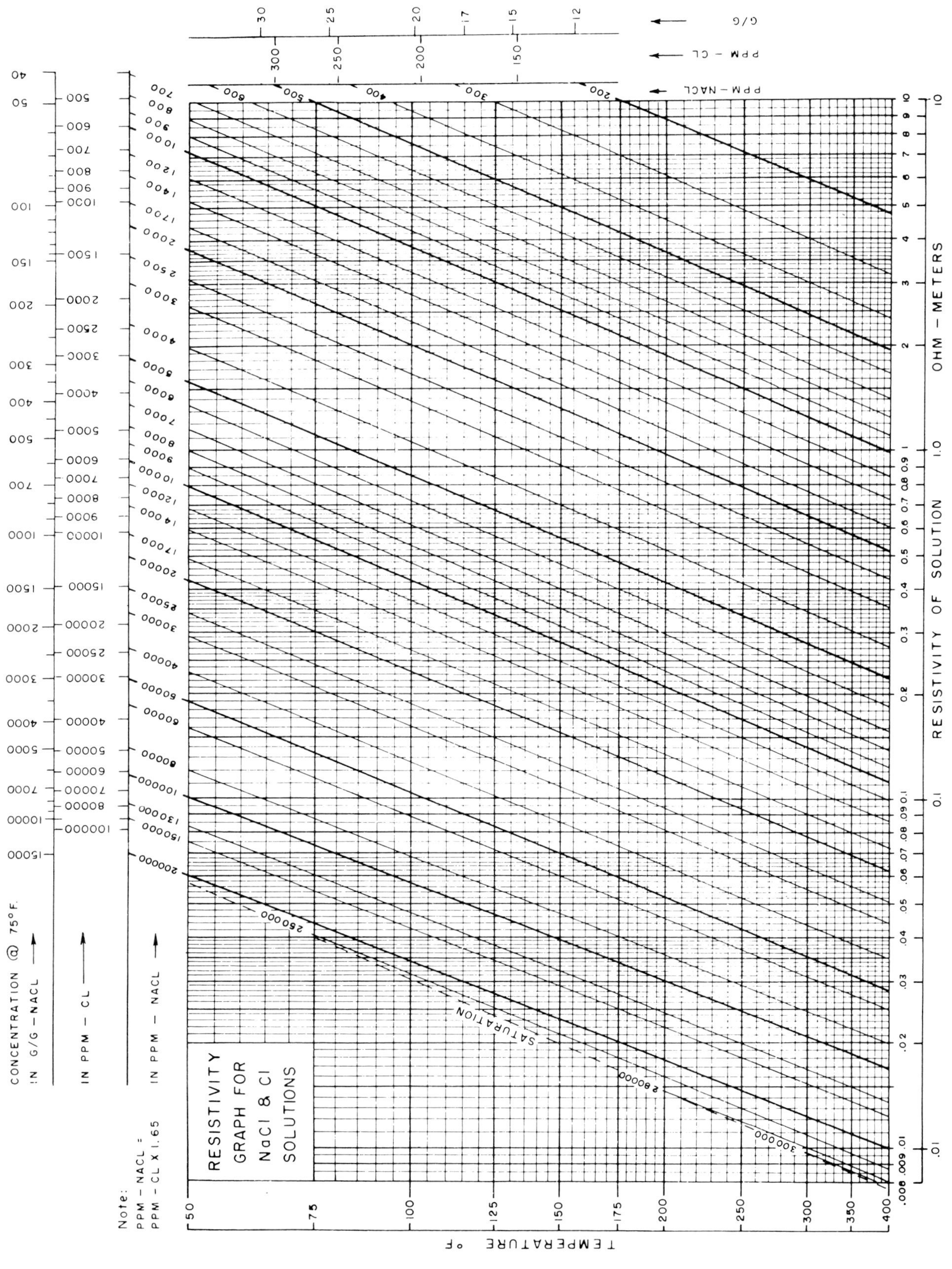
RESISTIVITY GRAPH FOR NaCl & Cl SOLUTIONS
Note: PPM - NACL = PPM - CL X 1.65
CONCENTRATION @ 75°F.
IN G/G - NACL
IN PPM - CL
IN PPM - NACL
G/G
PPM - CL
PPM - NACL
SATURATION
TEMPERATURE °F
RESISTIVITY OF SOLUTION OHM - METERS

Chart FT-3

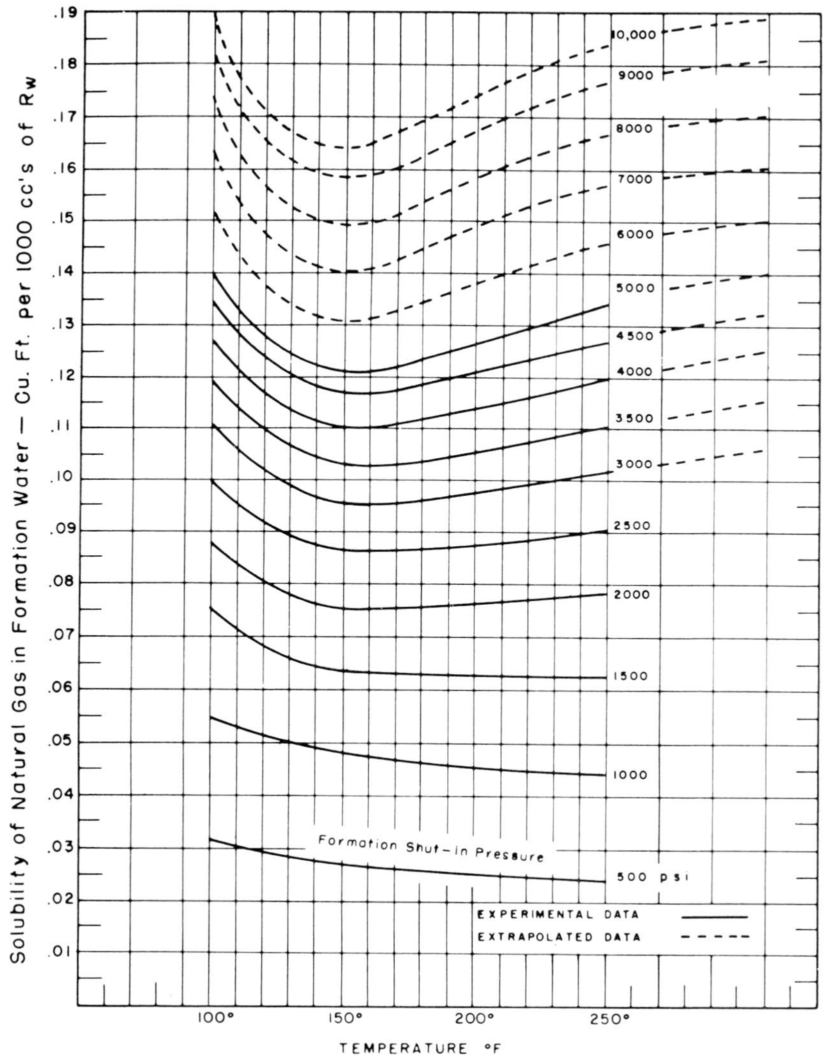


Chart FT-4

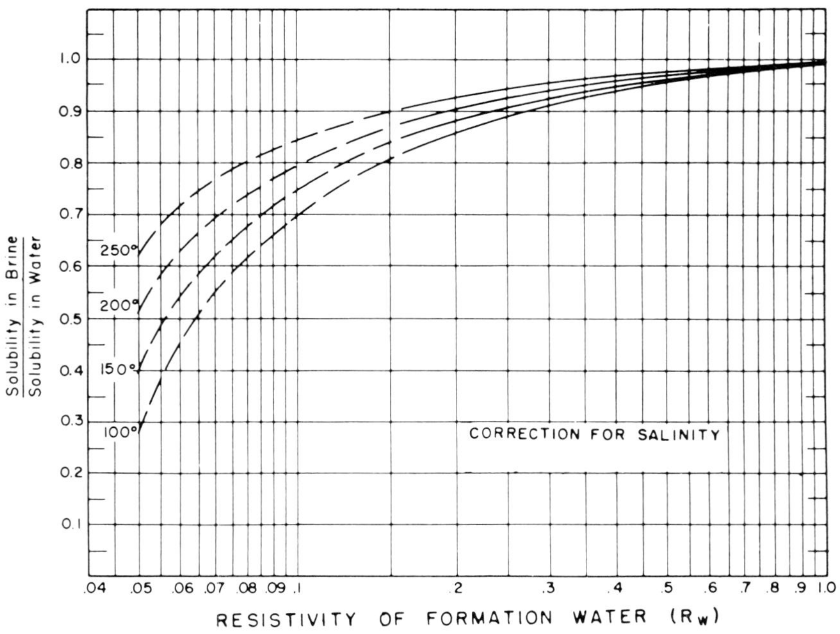


Chart FT-5

To determine the maximum volume of gas dissolved in the recovered formation water:

1. Enter FT-4 with formation temperature and formation shut-in pressure. The ordinate of the intercept defines the maximum volume (cu ft) of gas soluble in 1000 cc of pure water.

2. Correct for water salinity using FT-5. The intercept described by $R_W$ (at formation temperature) and formation temperature gives the salinity correction factor.

3. Multiply the volume of recovered formation water by the factors determined from both FT-4 and FT-5.

Charts FT-4 and FT-5 are adapted from data presented by Dodson and Standing, API Drilling and Production Practice, 1944.

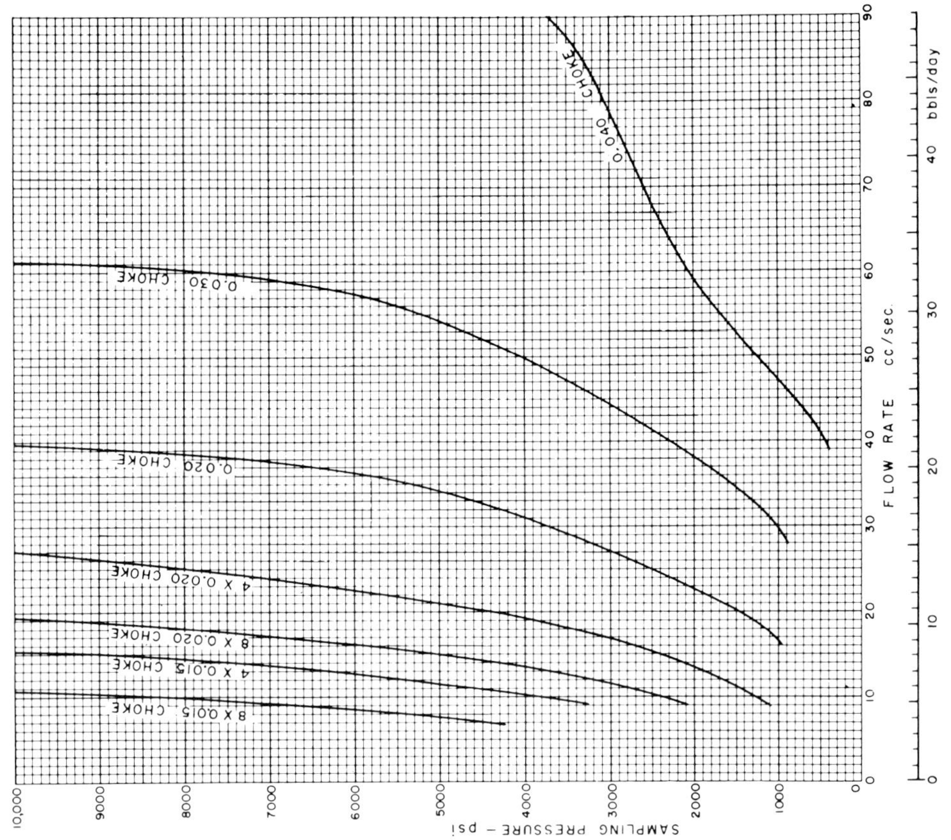


Chart FT-7

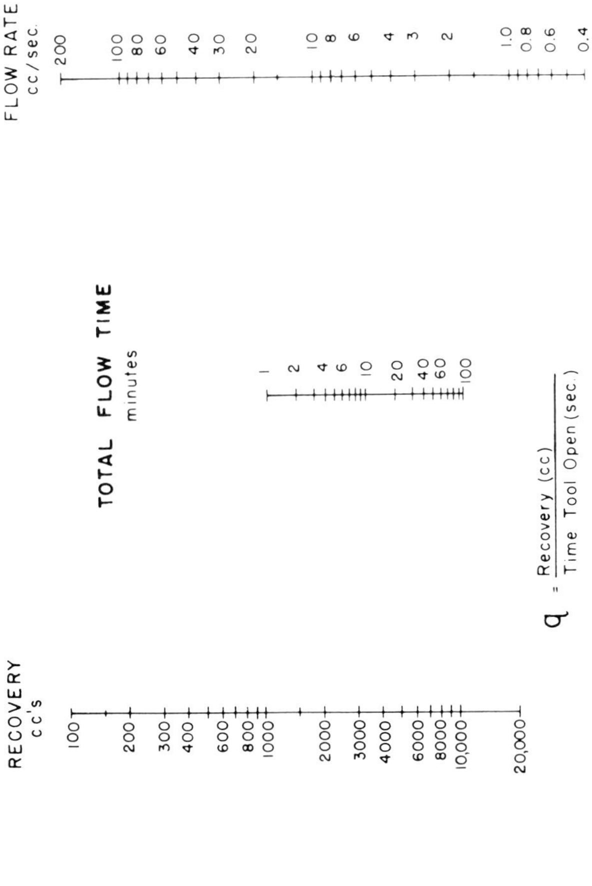


Chart FT-6

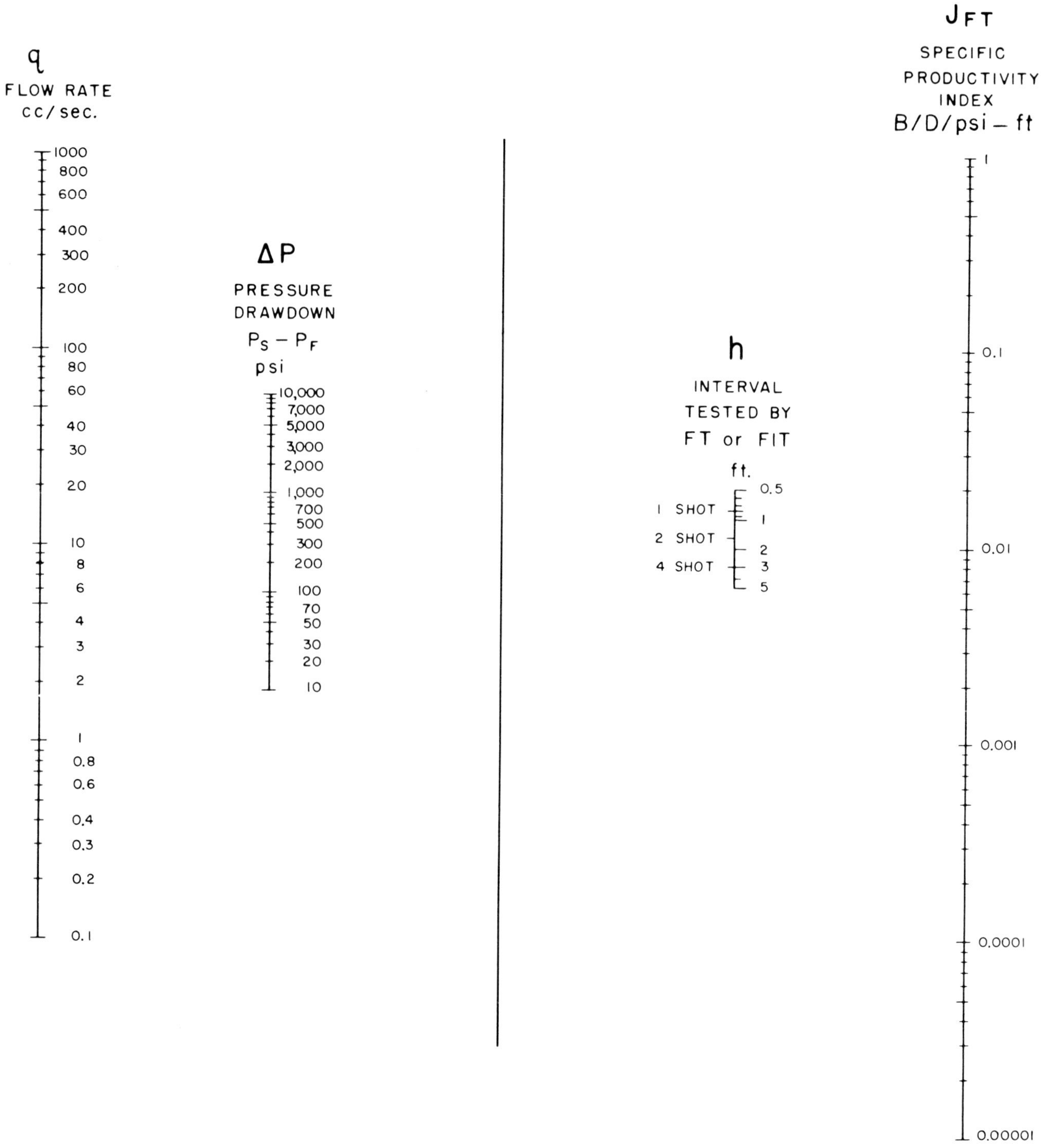
q
FLOW RATE
cc/sec.
1000
800
600
400
300
200
100
80
60
40
30
20
10
8
6
4
3
2
1
0.8
0.6
0.4
0.3
0.2
0.1
ΔP
PRESSURE
DRAWDOWN
Ps − PF
psi
10,000
7,000
5,000
3,000
2,000
1,000
700
500
300
200
100
70
50
30
20
10
h
INTERVAL
TESTED BY
FT or FIT
ft.
0.5
1 SHOT
1
2 SHOT
2
4 SHOT
3
5
JFT
SPECIFIC
PRODUCTIVITY
INDEX
B/D/psi − ft
1
0.1
0.01
0.001
0.0001
0.00001

Chart FT-8

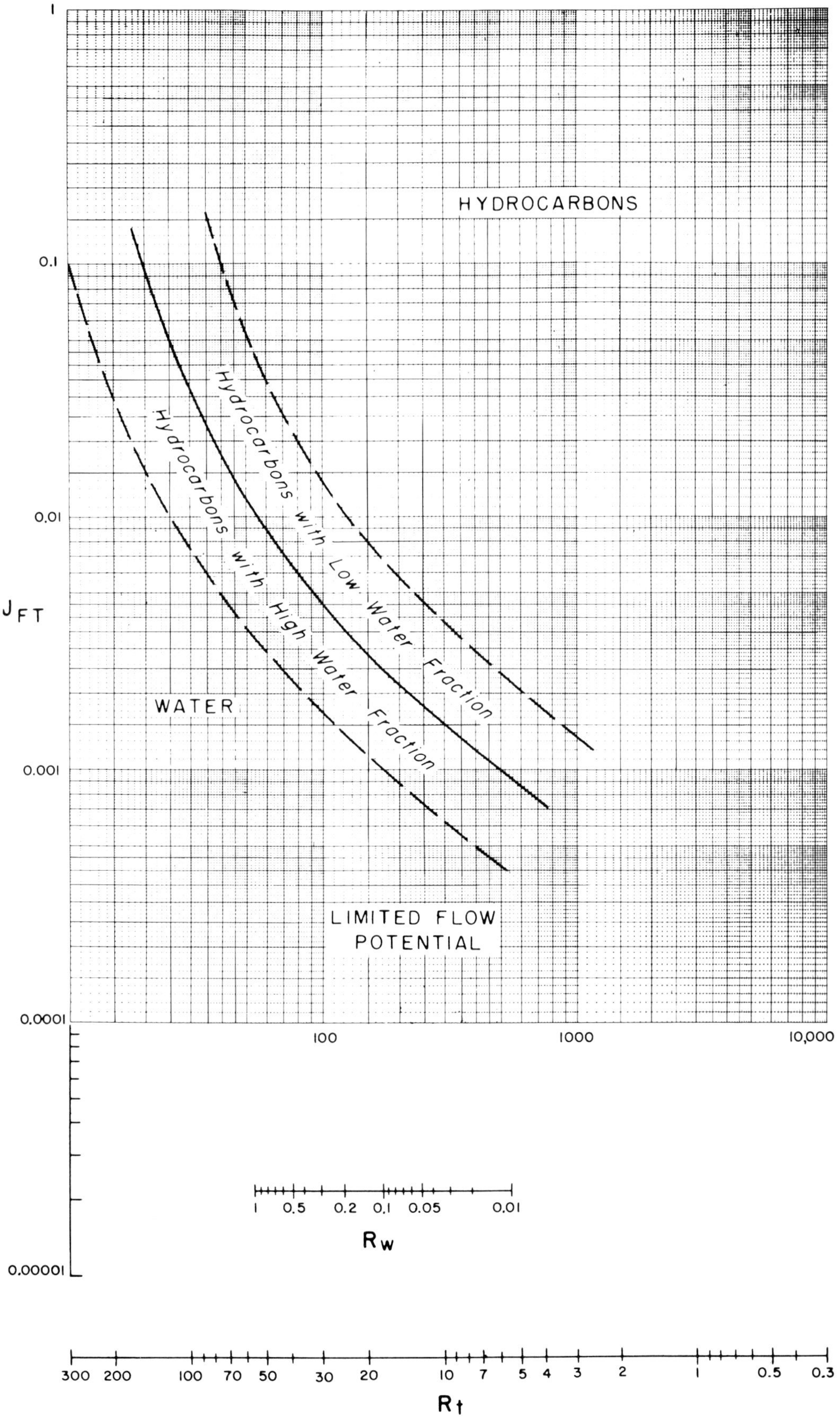
HYDROCARBONS
Hydrocarbons with Low Water Fraction
Hydrocarbons with High Water Fraction
WATER
LIMITED FLOW POTENTIAL
$J_{FT}$
1
0.1
0.01
0.001
0.0001
0.00001
100
1000
10,000
1 0.5 0.2 0.1 0.05 0.01
$R_w$
300 200 100 70 50 30 20 10 7 5 4 3 2 1 0.5 0.3
$R_t$

Chart FT-9